Quantum Markets

Physical Theory of Market Microstructure

First edition

Jack Sarkissian

(2020)

Jacob Sarkissian

New York, NY USA

ISBN 978-0-578-79637-6

Advanced Scientific Publishing, 1110 Nuuanu Ave, Honolulu, HI 96817

Printed in the United States of America

Contents

Notations

τ – time between trades

ρ – order size density

θ – order density flow

Δ – spread

$\epsilon = \frac{\Delta}{s}$ – relative spread

h – height of high-low bars, extension of spread

I – execution imbalance

J – order imbalance

σ_e – ensemble volatility index

ρ_e – ensemble correlation index

T – time horizon

p – probability

ψ – probability amplitude

s – price

x – log-price

m – order/trade size (absolute value)

q – order size (signed), executing order size, child order size

M – trade size (absolute value), total unsplit order size

Q – trade size (signed), total unsplit order size

V – trading volume

v – order flow

σ_x – standard deviation of variable x, unless context dictates a different meaning

κ' – real part of κ

κ'' – imaginary part of κ

$1\ bp = \frac{1}{100}\ of\ 1\%$ (basis point)

L – price impact

PREFACE

Many years ago, when I started working in finance, I thought it would be easy. I was starting as a risk analyst, which essentially involved computing portfolio statistics and interpreting it, so with a leverage like physics PhD I felt confident that it won't be hard. My expectation was that I would build risk models, plug in the data, and generate beautiful reports to send to the management.

That expectation didn't last long – just until I saw the portfolios. These were large portfolios, made of obscure securities resembling nothing of textbook exercises. Most securities traded irregularly, sometimes once a week or once a month, some never traded, so there were large gaps in the data. Even one such security would bring down reliability of the entire risk model. So, the question was, how do you build a risk model when there are big gaps in the data? Textbooks had no answers.

Risk models from textbooks predicted zero loss if you bought a security and immediately sold it. That obviously was incorrect, since to sell immediately you would have to sell at a discount and take a loss.

That's when I started to realize that something was essentially wrong with finance theory and that "something" had a lot to do with liquidity. My subsequent work on trading desk only confirmed this observation. Whenever liquidity issues came along, finance theory failed.

In the meantime, business didn't wait. How do you compute the bid-ask spread for market making? How do you price an oversize lot in an obscure security for a trade? How do you price it for a repo transaction? How do you control risk of illiquid securities and how do you price derivatives on them? Inability to get reliable results consistent with data and experience was turning my business into a game of guesswork in which things relied on "analyst estimates" and "expert opinion". To turn it into a business I needed a framework that overcame these difficulties.

That's what prompted me to look for solutions outside of traditional finance, eventually leading to development of quantum theory of price formation. Using physical approach on microscopy of trading process and questioning the basic concepts – everything pointed in the direction of quantum nature of price formation. As it developed more and more, I had to realize that it's not just a model – it's a physical theory with its concepts and foundation that affects the way we should think about financial markets. That material makes up the content of this book.

To be clear, the purpose of this book isn't in providing philosophical concepts. Its purpose is to provide a framework that will allow quantitative analysts, traders, and risk managers to solve practical problems commonly arising on trading desk and do it in a way that's applicable to institutional reality in which they work. The book is intended to help the reader understand the physics of underlying processes, put it in equations and obtain results that are valid for institutional scale.

This book deliberately does not contain theorems and proofs that are useless on trading desk. To quote Landau and Lifshitz from their famous 10-volume "Theoretical Physics" course – "we deliberately didn't pursue mathematical rigor, which in general isn't well achievable in theoretical physics". This does not mean that scientific rigor is compromised, but only that overcomplication of material is avoided.

However, one thing I kept strict about – it's the derivation of quantum equations as they apply to finance. Coming from probabilistic nature of finance and quantum mechanics some researchers made attempts to simply borrow quantum mechanical equations and draw their analogy with finance. That is incomplete. To do it right, quantum mechanical equations have to be REDERIVED for finance as they apply to finance, not elementary particles. This is true even if the resulting equations come out the same. "Borrowing" or "bolting" equations from quantum mechanics onto finance just based on a similarity can lead to a distorted picture, which is opposite to what this book is intended to achieve.

The way this book is written, it implies that the reader already has knowledge of quantum mechanics and statistical physics and is familiar with methods of theoretical physics. This will change with time as we find ways to explain these concepts to a broader audience – just the way it happened decades ago with stochastic calculus.

Unlike most textbooks that go from general into specifics I decided to develop the material from specifics to general. I believe such order will be more comfortable for the reader. Thus, the coupled-wave model is introduced in PART 2 before the continuous model. The coupled-wave model deals with the top levels of the book, which is where most of price formation takes place. Many notions have intuitive meaning in that model, so it's a better point to start.

PART 3 expands the coupled-wave model into general multilevel and continuous representations. It introduces the general equation governing evolution of security, which turns different from the Schrodinger equation. It combines both spread and diffusion process in a single framework evolving together and linked to the trading environment. To avoid confusion with Schrodinger equation, that equation is dubbed Market Maker's Equation.

Coupled-wave model's main achievement is that it allows to model securities as two-level systems. This opens a new dimension in studying their statistical properties through an ensemble view.

That path is taken in PART 4. As a method it allows to build fast reacting measures of current volatility and correlations, and bypass difficult micro-modeling of execution probabilities.

Some readers may be looking for material related to price prediction. There is none here. As someone who has been in business long enough, I know that price prediction borders on astrology. I count that the readers share that experience with me and instead of trying to predict future price movements focus on using this material for building better pricing, execution, and risk models.

Some authors may be disappointed not to find references to their work here. This book is not a review, and I don't feel obligated to include everything. I reference material that is useful on the trading desk. Abstract material, mathematical theorems and proofs, material based on unmeasurable or unquantifiable notions, as well as material that cannot be used to produce valid results or contradicts the scientific method is not referenced.

This book is built around the equity class, although the same concepts can be applied to any. Certainly, universal properties of price formation do not depend on particulars of company fundamentals, market conjuncture, etc.

As speculative as they are, today's financial markets are developed enough to be considered physical systems. I hope this book will serve not only as a technical guide, but also as an inspiration for continued research in this direction to uncover the beauty of seemingly the most chaotic world around us.

October 2020

Jack Sarkissian

New York, USA

ACKNOWLEDGEMENTS

Little I knew in the beginning of my finance career 15 years ago that a day would come that all the boardrooms, presentations, swap valuations, broker reports, analytic notes, ratings meetings, lunches with clients, and audits – everything that makes up a finance career – would convert into a beautiful picture that opens up when you look at the markets with the eyes of a physicist and which is described in this book. Most people would reserve it for commercial purposes, but I feel that at this point this knowledge does not belong solely to me and that it must be shared with the rest of the world.

And so, my acknowledgements go to all the professionals who appreciate the direction of my research and have continuously reaffirmed the need for it on a broader basis, particularly Joel Sebold, Revant Nayar, Olga Inglis, Igor Halperin, Vadim Nastasiuk, David Orrell, and Nikolai Zaitsev.

Experience that went into this book wouldn't have been accumulated without participation of my colleagues, clients, and counterparties. I must note Anatoly Predtechensky, whose professionalism made it possible for me to work in a rare high-standard environment, and Dr. Roman Maszczyk, who first gave me a green light into 10+ digit finance 15 years ago.

I continue to be inspired by my followers, younger professionals who question what they learn and keep digging it deeper for the correct answers as well as by those driven by natural intellectual curiosity.

I would like to thank my father who at my early age realized the need to introduce me to physics, who routinely took me to his two-floor Mossbauer effect lab to observe experiments and who answered my zillion questions about how nature works.

Lastly, I would like to thank Dr. Evgeny Margulis whose critical comments helped me structure the material, and Erica Wells who initiated the idea of writing this book.

In older days scientists looked for a unified picture of the world and called it in Latin “congruit universa”. There is much more work to do, but at this point I think it is safe to say that we are on the right path to have “congruit universa” in finance.

INTRODUCTION

Difficulties of quantitative finance

Decades ago, only select institutions or individuals were able to get access to trading. Today, trading has become so accessible that pretty much everyone can do it. With better accessibility and bigger competition interest has now shifted from "how to trade" to "what price do I get when I trade". This statement defines the problem of trade execution.

Prompted by rapid development, pushed by the lowering commissions and financing costs, and enhanced by the regulations – MIFID II in Europe and the upcoming Reg NMS II in the US – question of trade execution is becoming increasingly important. Answer to that question lies in study of market microstructure and primarily, price formation.

Application of mathematics was able to achieve much success by introducing stochastic calculus to finance. It progressed our abilities to price derivatives and fixed income instruments, model investment portfolios and evaluate risk. Inspired by this success it is now being applied to execution.

The truth is that stochastics cannot handle execution. As we will see in the following chapters, its power is limited by very liquid instruments that do not really pose an execution problem. Every mathematical model has its limitations and stochastics reaches its limits at market microstructural level just like classical mechanics reaches its limit at quantum scale. And just like classical mechanics runs into paradoxes when it's applied on quantum scale, so does the stochastics.

A popular paradox is the risk of loss at zero time horizon (think about zero time horizon as non-zero, but very small). Because Value-at-Risk scales at $\sqrt{T}$, it appears as if losses at $T = 0$ are also zero. It's amazing how many risk managers give that result for an answer. The right answer is of course that buying and selling prices are different, so if something was just bought, it will have to be sold cheaper. This difference between buying and selling prices is not part of stochastic calculus, and therefore the result obtained with it contradicts observations.

Mathematical models define price as the expected value of associated cash flows. The result is that buying and selling prices are the same, and that price for a 100 and 10,000 unit lots are also the same, which again contradicts observations. These differences are introduced afterwards, by applying econometric adjustments or through traders' intuition.

As an illustration, let's look at some of the problems that arise on the trading desk and where the classical stochastic framework meets with serious difficulties.

Market making

Market makers are firms or desks within firms that quote buy and sell prices of financial instruments for other market participants so they can have these prices readily available – much like currency exchange shops at the airports. Market makers profit from bid-ask spread, so spread management is crucial to them. Because of that, they compete for larger turnover in quoted securities.

Higher turnover can be easily achieved by reducing the operating spread – the difference between market maker's own buy and sell quotes. On the other hand, doing so also lowers profit per trading roundtrip. To the contrary, increasing the operating spread increases profit per cycle, but reduces turnover.

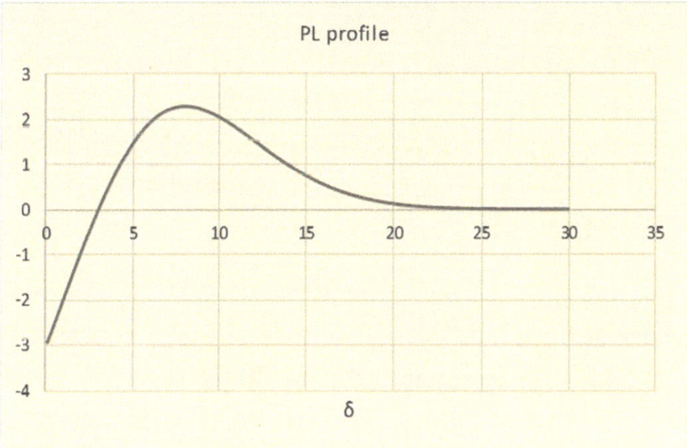

FIGURE 0.1 *Tradeoff between market maker's operating spread and P/L. Profit per cycle grows as spread increases but starts to shrink after passing the optimal spread.*

Apparently in the two extremes – very small spread and very large spread – profit vanishes, but is positive in between, as shown in FIGURE 0.1. Therefore, it must reach the maximum somewhere in the middle. Spread that maximizes P/L is the optimal operating spread and the market maker has to determine it.

To be able to answer that question we need to know how turnover depends on spread and how market conditions affect it. With this information at hand, we can express the P/L as a function of spread size and optimize it. Stochastic calculus has no way to approach this problem because the notion of spread is missing from it.

Sales-and-Trading (S&T)

S&T departments in investment banks buy and sell securities on behalf of the bank or its clients. As orders come in large sizes, execution quality is particularly important.

Academics view optimal execution as a problem of slicing orders in time to minimize price impact. Viewed from an S&T perspective, execution has many more facets. Traders work with the order book and other venues to keep the spread minimal and maximize filling probability. Most of execution process is focused on these capabilities.

As orders are submitted, the order book shuffles reacting to the new arrivals. Being able to anticipate market's reaction **in its depth**, being able to spot hidden liquidity and differentiate between genuine and fleeting liquidity are important elements of the process.

These are examples of what order execution is really about and not just slicing and dicing orders in time to minimize market reaction.

REPO transactions

Look at a situation when a client wants to REPO a stock (stock!) of a gold-mining company in which he owns 37%. He is looking at a 2-week rolling term for a large block. The stock is traded with medium volume around 400K/day and has a 34% annual volatility.

Usually, REPO transactions are made using government bonds as the safest and liquid instruments. Corporate bonds involve bigger risk but are also quite popular in REPO. Single name equity REPOs are rare and usually come from individuals or firms that are already clients for other services, but cases like this one provide lucrative opportunities.

The issue with equities is the big price variation, even over the 2-week period. This substantially complicates haircut calculation. The usual VaR or Expected Shortfall based calculation is insufficient, because in case of nonpayment the equity block will have to be liquidated in a thin market which substantially adds to haircuts. With that in mind, you have to propose terms that will put you out of risk, but also be acceptable for the client. One more case, where stochastic calculus fails.

Trading

A short-term equity position finally hits positive P/L, according to the trading terminal. Trader decides it's time to close the position and take profit. He submits a liquidating order. The position represents 10% of the daily volume, so liquidation takes a while. By the time it's finished trader discovers that instead of fixing a profit he took a loss.

Trading terminals assess P/L based on current market prices. Those prices are based on 100-200 share lots that usually go through the exchange. As position liquidates it affects the market bringing the price down and reducing the effective sale price.

It's important that position valuation be based on realistic expectations that are adequate for position size, even if an active but shallow market exists. Same relates to investment performance reporting, particularly in hedge funds and pension funds. Trade size must be a natural pricing input in realization of the IFRS principles.

Asset management

Another situation: a trading counterparty knows that you have a certain bond in portfolios, and they want to purchase a block that amounts to 25% of the issue. They contact you to discuss the terms. The last trade of comparable size occurred 2 months ago, but there are small transactions every few days and the bond has liquidity in the book totaling about 3% of the block size. Interest rate market allows a variation of ±50 bp from each side. You also know that there are only 3 other firms with substantial block sizes. How do you price the block and convince the other party that it's a fair price?

Stochastic models give a fair price without regard to size. In that sense, they propose trading a $60 mln. block at the same price as a $6,000 lot. Eventually, price ends up being adjusted for size using econometric modeling that explains nothing or by intuitive expertise. In a solid quantitative framework security price must come as a function of trade size naturally without artificial adjustments.

Risk management

Many risk departments take their job as simple reporting. It's popular to think that as long as they have computed probability distributions, portfolio statistics and produced nice reports their job is done. Meanwhile, the real purpose of a risk department is to PREVENT losses. When losses do happen, their magnitude usually prevails over the estimates.

Just like in the previous paragraphs, an important factor contributing to this is that risk evaluation is being based on prices that are valid for small lots going through the exchanges. Evaluation that disregards market impact as the positions are liquidated will understate risk. As a result, realized losses can be much bigger than expected.

One of the tricks used by risk managers to help evaluate risk of illiquid securities in the portfolio is gap filling. Gaps in trading data are artificially filled and then used along with the liquids to compute VaR and shortfall. Instead of reporting the liquidity structure of the portfolio and realistically estimating the potential losses arising not only from price fluctuations, but also from price impact, data is being distorted to facilitate a convenient computation, which does not help prevent losses.

Options on illiquid securities

Illiquid underlying complicates the already complex picture of option pricing. Apart from the unknown dynamics of the underlying, benefits of the option holder now become size dependent. Additionally, having an uncertain price translates into uncertainty of the Greeks and complicates the hedging process.

Modeling such options is possible only through modeling illiquid underlying with different bid and ask prices and the right statistics of spread, which again extends beyond the scope of stochastics.

Methodological aspects

Difficulties listed in the previous section have one thing in common: they arise from the lack of ability to relate price to liquidity. That lack is an artefact of descriptive approach which aggregates trading data into statistical measures but doesn't connect them further to the building blocks of trading.

Markets aren't made of probability distributions, volatility surfaces, zero curves, etc. These notions are certainly helpful, but they are not, figuratively speaking, on the elemental chart. Before any trade happens, there must be quotes and orders preceding it. These are the primary elements, whose interaction with each other leads to trades. Market microstructure studies how that happens.

The ability to relate microstructural elements to trading results allows to solve the problems listed in previous section. Since this book describes a physical theory, let's look at the current state of finance theory from the perspective of physics.

In every field of physics there is a general, central equation or set of equations that solve everything in that field by applying them to specific problems. For example, Einstein's Field Equations in General Relativity relate the geometry of space-time to distribution of matter in it. All the particular problems are solved by applying these equations to specific cases. Likewise, Maxwell's equations form the basis of electrodynamics, Boltzmann equation is the basis of kinetics, etc.

In finance models are produced in bottom-up direction, by starting with basic cases and introducing addons. For instance, the jump-diffusion option pricing model was developed by adding jumps to Brownian motion process of price. Each step up brings with it one or more postulates. In this way the postulates pile up leading to eventual loss of the internal natural logic. Deriving the general equation from the basic principles, an equation that allows to plug in microstructural parameters and obtain observable results is one of the goals in study of market microstructure.

Certainly, level of fluctuations in finance doesn't help the discovery of the general equation. To this extent, numerous interpretations often can fit the same observational set, but there is only one that fits all phenomena simultaneously.

Feasibility of models and theories can be established only on the basis of measurement. Models that juggle notions that are, albeit intuitively understandable cannot be measured, do not belong in quantitative finance. They can be related to economics or sociology, but not to quantitative finance. Lack of verification for feasibility is what gives us models that are so abstract that they border on fantasy. That's how we get models of pension funds that are funded by loans[1], market making strategies that require daily liquidation of inventory[2], and execution optimization with linear impact and market orders[3].

Association of finance with human behavior made it to be perceived as a social science. That would be okay, except such perception brings with it its lowered standards for measurement and hypothesis testing. The result is that finance researchers are often satisfied by a mere similarity of behavior, and not by correspondence of exact functional behavior. From that perspective $E = mc$, $E = mc^2$ and $E = mc^3$ are the same equation. Numerous models are being produced based on a mere behavioral similarity without regard to the ability to measure and test. Here are a couple of examples.

It has become popular to bring up Heisenberg's uncertainty principle in finance. A qualitative statement that there is an uncertainty relation between price and volume without producing and demonstrating that relationship is insufficient. Uncertainty relation is not a philosophical concept, but a very specific inequality relating measurable quantities to each other.

---

[1] Pension funds are funded by earnings, not loans.

[2] Market makers can't liquidate every day because they are the ones that make markets. If they did liquidate, they would return all day's earnings and pay extra.

[3] Market orders are the straight way to getting the worst price and therefore defy the purpose of execution. Additionally, linear price impact occurs in heavy volume trading, which also contradicts the purpose of execution.

Same is true about the nature of market phenomena. It has been long speculated that market crashes are connected to second order phase transitions. Proposed models simply repeat their well-known counterparts from physics (Ising model, etc.) without defining measurable input or providing experimental evidence. Phase transitions are not a sociological concept and are characterized by very distinct physical behavior. Unless this behavior is confirmed, one can only relate to hypothetical resemblance between the two processes.

Despite the opposite approach, abstract mathematics often achieves a similar result. Abstract mathematical approach puts models above the markets. Usually, it is able to describe market behavior in some region of parameter space, but agreement with observation isn't fully verified as logical reason and rationale are considered to be sufficient for claiming model validity. If market behavior deviates from mathematical predictions, then the deviation is resolved by declaring markets inefficient. By putting models above the real market behavior abstract mathematical approach discontinues its relevancy the moment it invokes the idea of market inefficiency.

Physics aims to study the world around us. The world, and not logic, reason, similarity, or analogy is the central object in physics. In physics every variable has a meaning and can be measured directly or computed from measured data. And if the observed behavior deviates from the model, the model is discarded no matter how logical or reasonable it seems, to be replaced by a different model that fits the reality. With that in mind, let's look at the main differences that physics approach brings into finance as compared to traditional mathematical, econometric, and sociological ones.

Why physics?

Financial markets have developed enough to establish themselves as valid physical systems. As long as physics is a discipline that studies the world around us with the goal to understand how the universe works – financial markets are a perfect object for such study.

Depth of understanding. Physics connects every model variable to observable and measurable natural quantities that have physical meaning. Interrelation of different variables in a meaningful and observable way brings with it a deeper level of understanding. Consider an example: what would happen if some of the modern analytic techniques were used hundreds of years ago to study the motion of the planets.

Say, someone wanted to build a model predicting the positions of planets on the sky. They could, potentially, build a deep learning model and train it on the existing past data. The result would be a complex model with a host of parameters, that forecasts planetary positions on the sky days, weeks, and months into the future. In the meantime, the entire model is just sitting on a simple algebraic law that is the Newton's Gravity Law with a single parameter – the gravitational constant. Accuracy of the deep learning model will deteriorate with time, whilst the Gravity Law has been around for ages.

Since the task is to forecast positions on the sky and not in absolute space, we wouldn't know that they orbit the sun and make up the solar system. We would, of course, miss Kepler's 3 laws of planetary motion, and the Gravity Law that stands behind them.

Worse yet, we wouldn't have discovered Neptune and Pluto. After discovery of Uranus using advances in celestial mechanics astronomers started making tables of its future positions. It was noticed that the trajectory of Uranus deviated from calculations. Assuming that another planet's gravity is causing these deviations, that planet's orbit and position on the sky were calculated. Neptune was discovered only $52'$ away from the calculated spot. Around a century later Pluto was discovered in a similar way. Projecting this onto modern day quantitative finance, positional deviations could have been simply written off as "inefficiency of the Gravity Law".

This being said, machine learning and artificial intelligence cannot be entirely replaced by physics approach. There are numerous instances in finance when, albeit the relationship is apparent, building a physical model is so convoluted that it's not worth the effort. In situations like these, physical and mathematical approaches are complimentary.

No overfitting problem. Physics doesn't have some of the problems mathematical finance has. Overfitting is a great example. Overfitting occurs when model has too many parameters. With the extra parameters at hand, model achieves a better fit with data because more parameters can be tweaked. However, as soon as the model moves to new data it breaks down. Overfitting is dealt with using optimization methods that allow to select the important parameters and discard the unimportant ones.

Physics doesn't have an overfitting problem. In physics equations come out ready for use. Each quantity has its place in the equation and there is no question about importance of parameters. For example, the Gravity Law has only one parameter – the gravitational constant and it's a multiplier in the formula. Likewise, equation $E = mc^2$ has one parameter – the speed of light. All that's left once we have the formula is to determine parameter values from experiments.

As physical equations are applied to different problems, some terms may become less important by order of magnitude. Then the unimportant terms can be dropped to simplify equations. But preserving them has no negative effect on the end result and should not be confused with overfitting.

Model limitations. Every model comes with its limitations. For example, geometrical optics is valid when wavelength is small compared to objects, and Newtonian mechanics is valid at low speeds. Physics recognizes this property of models and allows to keep track of applicability limits before applying the model.

Generally, question of model limitations is not asked in finance. Rather, financial models focus on advantages and drawbacks for a particular use. Usually, model limitations are discovered only after large trading losses. But even then, some (if not many) quants blame the markets by claiming that "the model is good, but the markets changed".

Physical model construction allows to trace postulates and assumptions built into the model, so that they can be verified before applying it. For instance, the coupled-wave model described in PART 2 is valid when trading flow is small enough not to pierce through the top levels of the order book. We are aware of this limitation before applying the model and so we can always check that the applicability conditions are satisfied to be sure to get valid results.

Beauty of equations. Unlike equations of abstract mathematics, physical equations tend to have certain quality, that is intuitively described as beauty. This quality involves equational symmetry, dimensional units, asymptotic behavior, polar structure, etc., that reflect the fundamental properties of physical systems. While there can't be a proof to such statement, equations that do not possess this quality cannot describe the nature.

For instance, volatility is an unnatural quantity. Its dimensions are $\frac{1}{\sqrt{time}}$. We realize that there isn't a process in nature, that involves square root of time. If volatility appears in equations, it can only be in form of its square: σ^2. The only time when volatility appears in equations linearly is when the function's square depends on square of volatility, $y^2 \sim \sigma^2$, and therefore taking a square root creates an impression of linear dependence.

Inner beauty of physical equations is like a litmus test that helps distinguish real models from artificial ones.

* * *

How does price form? What determines the next trade? What happens between trades? How does market react to a trade or to an order before trade? What makes securities liquid? And what equation describes evolution of securities starting with microparameters? In the following parts, we will try to answer these questions from physical perspective.

PART I

PRICE FORMATION MECHANISM

1.1 Ambiguity of price

It is customary to reference security prices as single numbers. For example, someone might say that price of Citigroup Inc. stock (ticker "C") on April 18, 2016 was $45.11. That was indeed Citigroup stock's closing price on that day. While good enough for many uses, closing price is not the only choice. For example, one may reference the price of the last trade instead. Bid price can be referenced for sellers, asking price for buyers, or both prices can be referenced for traders to "show them the market". There is usually a variety of options to reference a security price. This variety is common to institutional finance and is not accidental.

In financial markets price forms as a result of interaction between buyers and sellers. The most widely known venue in which this interaction takes place is the securities exchange. An exchange combines buying and selling orders into a table called **order book**. An example of order book is shown in FIGURE 1.1.

In this example buyers are buying at lower prices and are shown on the left side, and sellers are selling at higher prices and are shown on the right side. The trading parties wait in line for a matching order. Matching order can come as a newly submitted order, or as a modified existing order.

First candidates for matching are called the **top of the book** – the highest bid s_{bid} and the lowest ask s_{ask}, marked green and red in[4] FIGURE 1.1. They are spaced out by the distance which is called the **bid-ask spread**:

$$\Delta = s_{ask} - s_{bid} \tag{1.1}$$

Without a transaction we can say that security **price is localized between the best bid and the best ask**. When an order is matched it disappears from order book and a transaction is recorded. Then the security price becomes certain, but only within the context of a particular transaction with its timing and size. If a trade just occurred, another trade happening a moment later may transact at a different price.

Buyers – bid	
Price	Size
27.83	100
27.82	100
27.80	200
27.79	600
27.78	100

Sellers – ask	
Price	Size
27.87	100
27.90	100
27.95	1000
28.15	300
28.20	400

FIGURE 1.1 *Typical order book. Buyers (on the left side) want to buy the security at a lower price and sellers (on the right side) want to sell it at a higher price. Best bid is marked green, and best ask is marked red. These are called the top of the book.*

[4] Except when they come from the same party, but then that party should be aware of that.

Trade with different size will generally receive a different price. For example, larger buy orders will receive a higher price as the buyer has to pay a premium for extra quantity. Likewise, larger sell orders will end up being sold at discount.

This effect may seem counterintuitive since common mindset tells us that wholesale purchases must go at discounted prices. This is not how it works in financial markets. A larger purchase order signals interest in security. When sellers become aware of it, they react by moving their asking prices upwards. The result is that the market price moves in unbeneficial direction. Price premium provides a cushion for that move[5]. In the end, price variation with size will look like shown in FIGURE 1.2. The two branches in the chart correspond to buying and selling prices. Trade size is a major factor affecting price.

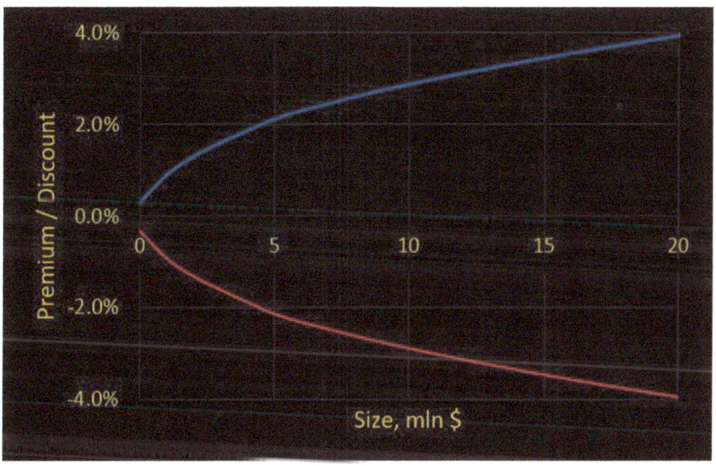

FIGURE 1.2. *Price dependence on trade size.*

---

[5] This should not be confused with the reducing scale of retail private banking service for smaller trades, which is simply intended to ensure bank's pay for smaller traders.

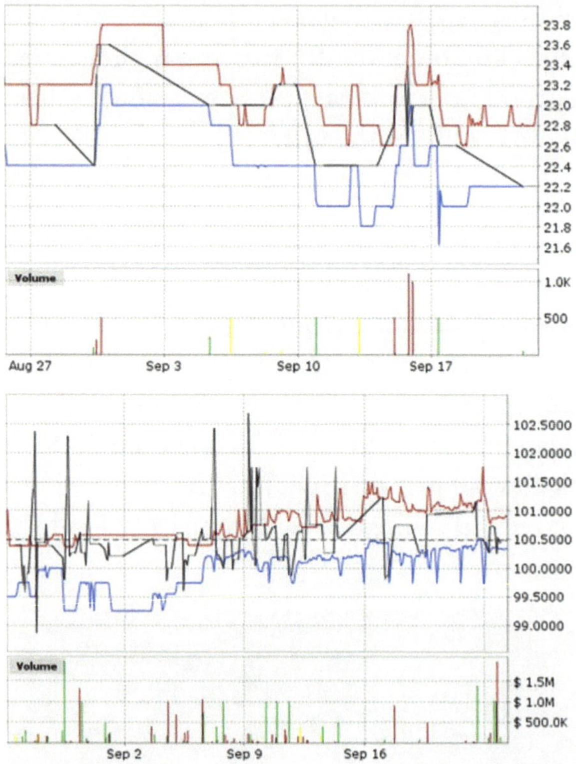

FIGURE 1.3 *Typical institutional class securities. Red line represents asking price, blue line represents buying price, black line represents last trade price. Left: FRIWO AG stock (CEA @ XETRA). Traded on exchange a few times a month and in very small quantities. Right: MPLX 6 7/8 bond, July 31, 2019 issue. Even though the security is traded every day, it is illiquid for an order size of over $1 mln.*

In institutional finance many securities do not have a current price. Numerous corporate bonds, municipal bonds, government bonds and derivatives fall into this category. Many securities are only traded a few times a day, many have not been traded for months, and many have never been traded before. An example of this is shown in FIGURE 1.3. In the meantime, it is these **illiquid securities** that make up the majority of institutional trading.

Quotes for illiquid securities have to be solicited from interested traders. Some securities lack liquidity so severely that one may have to spend a considerable effort and await weeks before being able to obtain the first quote.

One may think that such individualized price formation process is somehow different from the automated process on the exchanges. That is a wrong perception. Electronically executed trades are a result of preparation, negotiation and formation based on available information just as much as the individualized trades. The underlying nature of price formation in the two venues is the same, and therefore same are the laws governing price formation.

Price is an ambiguous, complex notion. In the following sections we will bring some order to the process of its formation and lay ground for mathematical description of it.

1.2 Limitations of stochastic calculus

Modern description of price dynamics is built around stochastic calculus treating price movements as random walk[6]. In doing so it implicitly relies on certain assumptions, that are supposed to ensure similarity between the two processes. Over the decades, countless models have been produced based on this similarity. To apply these models in practice in a valid way it is important to be aware what

[6] Standard references include (London, 2004), (J.P. Bouchaud, 2004), (Voit, 2001), (Jorion, 2006), (Hull, 2015)

criteria ensure similarity between price movements and random walk (Halperin, et al., 2020)[7].

Stochastic calculus assumes that security price is always available. Indeed, for any point in time suffice it to propagate the stochastic process up to that point and the price result is guaranteed. As it was pointed out, the reality is different. "Price availability" assumption may be suitable only for well traded, liquid securities, such as for example the blue-chip stocks, index futures or select issues of government bonds. Even then, frequent transactions do not guarantee price availability, because price availability for 200 shares does not automatically imply availability for an institutional level million-share block.

Another significant simplification of stochastic calculus is its treatment of price as a single number. We have already discussed price ambiguity in the previous section, particularly the difference between buying and selling prices. This difference is especially visible in institutional trading where transaction size is large. For example, it is common in fixed income markets to have the buying and selling prices set apart by 50 bp[8] or even 250 bp. The assumption that buying and selling prices are equal rests on implication that (a) securities are liquid (well-traded) and (b) are traded in small quantities. For such securities, bid-ask spread is small compared to price itself and possibly can be neglected:

$$\frac{\Delta}{S} \ll 1 \qquad (1.2)$$

---

[7] also (Derman, 2011), (Taleb, 2010), (Taleb, 2005)

[8] Basis point (bp) is 1/100-th of 1%, common measure of spread in finance.

Even then, some trading desks and firms, called market makers, live off this difference by turning securities over quickly and capturing the spread with high frequency.

The only time when it is valid to say that price is a single number is **at the time of transaction**. Transaction occurs when buyer and seller agree on price and exchange M units of security for a price $\$S$, thus reducing price to a single number. That case is special and deserves detailed consideration, which will be given in the next section.

Another case when the spread plays an insignificant role is when the time interval is so large that price difference between the initial and final values s_0 and s_T is large compared to the spread:

$$\frac{\Delta}{s_T - s_0} \ll 1 \tag{1.3}$$

These two inequalities, (1.2) and (1.3), express the limitations of stochastic calculus as applied to modeling price movements. These conditions are satisfied for small trades in liquid, well-traded securities. They are typical for small retail accounts, such as someone's individual brokerage account or a retirement account.

Conditions (1.2) and (1.3) are generally inapplicable for institutional level operations. Financial institutions cannot rely on analytics for billion-dollar portfolios that is based on prices relevant for only 200 shares worth only tens of thousands of dollars. It is impossible to rely on risk reports with a 10-day time horizon when it takes 3 months to liquidate the portfolio at a steep discount. Lastly, it is against common sense to use the same price valuation for buying and selling orders when the main point of institutional finance is in differentiating between the two[9].

To be applicable for institutional level operations, pricing framework must acknowledge the realities of price formation. It must recognize the fact that price is not always available and the fact that there can be two or more prices depending on trade direction and size. Theory of price dynamics must go beyond a mere analogy with random walk and be based on nature of involved processes, objective observation, and measurement.

1.3 Financial transaction as price measurement

The two assumptions discussed in previous section have in fact the same and deeper source: they both rely on traders' ability to quickly locate securities, agree on price, and make a transaction. What they miss is a realization that:

EVERY TRANSACTION IN FINANCIAL MARKETS IS AN ELEMENTARY ACT OF PRICE MEASUREMENT.

To see this, let us appeal to the definition of market price as it is commonly set in IFRS (International Financial Reporting Standards) and ISDA (International Swaps and Derivatives Association) documents (FASB, 2010), (ISDA, 2016). An asset has a certain market price when there are parties transacting the asset at

---

[9] In other words, stochastic framework does provide solutions, but those solutions only work for "spherical chicken in vacuum"

that price. From physics perspective a financial transaction is therefore a measurement act that puts a price tag on security.

Let us walk through the process of price "measurement" using order book from FIGURE 1.4. Here buy and sell parts of the book are combined with price in the middle column. To perform price measurement, we can submit a buy limit order for 100 shares at a test price to see if it is viable[10], FIGURE 1.5 (left). If the order does not fill, we will have to adjust the price upwards, FIGURE 1.5 (right). We will keep doing it until the order fills. At that point measurement is complete and the transaction price is taken as its result, in this case $27.83.

Ask size	Price	Bid size
100	28.20	
900	28.15	
100	27.95	
300	27.90	
100	27.87	
	27.83	100
	27.82	400
	27.80	100
	27.79	1200
	27.78	100
	27.75	300

FIGURE 1.4 *Sample order book. Red numbers represent orders to sell, green numbers represent orders to buy.*

[10] For clarity it is an idealized description. Actual process may be much more complex or volatile, but its nature stays the same.

Instead of buying, we could try selling as shown in FIGURE 1.6. In this case, we will have to keep lowering the price until order fills, which can happen at $27.87. The two measurements coming from the opposite sides produce different results, separated by the amount of spread. We can say that price has an uncertainty and spread measures the degree of that uncertainty. While localized within the spread price remains uncertain until the next transaction.

Ask size	Price	Bid size
100	28.20	
900	28.15	
100	27.95	
300	27.90	
100	27.87	
	27.83	100
	27.82	400
	27.80	200
	27.79	1200
	27.78	100
	27.75	300

Ask size	Price	Bid size
100	28.20	
900	28.15	
100	27.95	
300	27.90	
100	27.87	
	27.83	200
	27.82	400
	27.80	100
	27.79	1200
	27.78	100
	27.75	300

FIGURE 1.5 *Price measurement process coming from the buying side: (a) Buy order for 100 shares is initially submitted at $27.80 level (size adds to another already existing order). (b) To fill the order its price has to be moved up. Measurement result reads $27.83.*

Instead of using limit orders and gradually adjusting price, we could use market orders for immediate execution. Then the two measurement results would switch places, but it would still be the same two numbers and the same price uncertainty. This uncertainty is an intrinsic property of price and cannot be fully eliminated. It can be diminished by adding more liquidity to the market, but it cannot be fully taken away.

Ask size	Price	Bid size	Ask size	Price	Bid size
100	28.20		100	28.20	
1000	28.15		900	28.15	
100	27.95		100	27.95	
300	27.90		300	27.90	
100	27.87		200	27.87	
	27.83	100		27.83	100
	27.82	400		27.82	400
	27.80	100		27.80	100
	27.79	1200		27.79	1200
	27.78	100		27.78	100
	27.75	300		27.75	300

FIGURE 1.6 *Price measurement process coming from the selling side. The result reads $27.87.*

How do we improve price accuracy? A small trade may not have enough weight to represent price of the company share. Meanwhile, larger trade would have more weight, but impact the price. Here is how it happens.

In the experiment above we used 100 share orders. They hit the top levels of the book and stopped there. Larger orders may penetrate more levels. For example, market sell order for 700 shares would penetrate three bid levels and stop at the fourth with price $27.79. As a result, the 700 shares would sell for $27.814 per share, a $0.016 discount compared to the price of 100 shares ($27.83).

If one does not want to use market orders and places a sell limit order for 700 shares at a limit price above $27.83 intending to incrementally drop it, that order will not immediately fill. Moreover, the existing buyers may reposition their orders lower hoping to get a better price from larger seller. Additional execution mechanisms, such as iceberg orders, order hiding or chaining may reduce the impact, but it will still be present since the order's actual presence will alter the behavior of other traders over time. Either way, the result is that albeit carrying more weight a larger trade would distort price and increase uncertainty.

One may think that price accuracy can be improved by performing a sequence of measurements set apart by small time differential. For example, instead of submitting a 10,000-share bulk sell order, one could submit a series of 100 share orders. This may be effective in reducing price impact by allowing time for new liquidity arrival. Yet, this will not eliminate the bid-ask spread differential. Time will play against price accuracy as price drifts away under the influence of incoming news flow. The resulting trade prices will be, albeit more localized, but generally still not located on a smooth curve.

Let us summarize observations. We saw that generally price does not have a certain value. All that can be said about price – is that it is localized within and around the spread. Instead of certain value, there is a spectrum of prices that can be realized as a result of trade. Trade acts as price measurement momentarily bringing the price to a definite value. Meanwhile, measurement itself and its structure, can influence the outcome. Albeit localized, prices coming from different transactions will not sit on a smooth curve. **Price uncertainty is inherent to the process of price measurement and is rooted in its nature – it can be diminished but cannot be fully eliminated.**

These are not qualities of a classical measurement process (Dirac, 1982). They are qualities of a quantum measurement, and therefore **price formation process has quantum nature**. This conclusion comes as a result of observations and not as a result of a mere analogy. Consequently, mathematical framework describing price formation and its dynamics must also be quantum (Sarkissian, 2013).

Quantum framework replaces the notion of price with price operator (L.D. Landau, 1981). Price operator can be characterized by its spectrum, which contains all prices that the security can attain when it is transacted. Associated with each price level is the security's state. At each point in time security can exist in a superposition state with different prices until one of the prices is selected in trade, bringing the security to the state with that price.

Like the price operator, each quantity in quantum framework is represented by its own operator with its own spectrum. For example, operator that submits an order for execution, is the trade operator. And operator, whose expected value is the trade size, is the trade size operator.

Some operators commute with each other, others do not. In quantum mechanics' language it means that the corresponding quantities cannot be measured simultaneously with precise accuracy. This forms the basis for the uncertainty principle in finance.

One of the typical features of quantum world is interference pattern. Interference is washed away by the numerous interactions in which trades are executing randomly and with random parameters. Oscillatory patterns can be observed only in processes where trades sequentially amplify each other. Such are the critical events like market crashes, price corrections and rallies.

Compared to stochastic framework probability transfer in quantum framework is more diverse. In stochastics probability transfers due to its direct flow from one price level to another. That mechanism also exists in quantum framework. In addition to that, probability can attenuate at one level while at the same time amplifying at another. This gives rise to much richer variety of phenomena, such as balancing market reaction and collective behavior.

Despite its insufficiency at the micro level, stochastic framework has proven to provide satisfactory description of prices on a larger time scale when inequalities (1.2) or (1.3) are satisfied. Thus, quantum price formation theory must contain stochastics as its limiting case (Nastasiuk, 2014), (Nastasiuk, 2015). Practically speaking, quantum description must transform into stochastic (diffusion) framework when applied to securities with frequent transactions, deep liquidity, and dense order books (Sarkissian, 2013).

A few words must be said about the currently ongoing research that attempts to describe price formation as a mechanistic process by modeling order submission, modification, cancellation, and execution as sequence of classical processes. Expectation is that once these processes have been properly modeled all the market properties can be obtained by direct calculation.

Essential difference between classical and quantum descriptions has been extensively addressed in literature (L.D. Landau, 1981), (Dirac, 1982). It must be reminded that the fact that price formation bears features of a quantum process points that it cannot be fully described by purely mechanistic models, which attempt to handle behavior of individual traders. Models that treat price as a classical element may be applicable in limited cases but will fail if applied outside of their tight limits. This is why current attempts to understand spread's behavior within the stochastic calculus sooner or later lead to discrepancies and paradoxes. No matter how many extra variables and dimensions are added to the models, classical mechanistic approach will not provide a consistent description of price formation process.

1.4 Spread curve of price uncertainty

Price is not only uncertain at small time scale, but also at time scales that are much larger than time between trades. This uncertainty is commonly characterized with volatility of returns.

Volatility is defined as the standard deviation of returns:

$$\sigma_T = \sqrt{\frac{1}{N}\sum_{i=1}^{N}(r_i - \bar{r})^2} \tag{1.4}$$

where r_i is the return between time points i and $i-1$:

$$r_i = \frac{1}{T}\left(\frac{s_i}{s_{i-1}} - 1\right) \tag{1.5}$$

and $\bar{r} = \frac{1}{N}\sum_{i=1}^{N} r_i$ is the average return. Volatility involves time interval T and refers to that interval.

For a Gaussian random walk, if σ_{T_1} and σ_{T_2} are volatilities of returns based on intervals T_1 and T_2, they can be related to each other by the well-known square root law:

$$\sigma_{T_2} = \sqrt{\frac{T_2}{T_1}}\sigma_{T_1} \tag{1.6}$$

This law is a result of viewing price movements as a stochastic diffusive process. According to this formula, price accuracy can be indefinitely improved by reducing measurement time. This is not true in real markets since price accuracy cannot be lower than the spread[11].

This observation suggests that although volatility is a relevant measure of price uncertainty, it is incomplete. Another measure of uncertainty over time scale much larger than the time between trades is the height of **OHLC bars**. OHLC is the Open-High-Low-Close data based on trades aggregated in equidistant time intervals. "Open" is the opening price of the interval, "Close" is the closing price; highest priced trade makes up the "High" price and the lowest priced trade makes up the "Low" price.

The height of OHLC bars represents price uncertainty over its reference time interval, so the following mapping can be made:

$$s_{ask} \rightarrow s_{high}$$

$$s_{bid} \rightarrow s_{low}$$

$$\Delta = s_{ask} - s_{bid} \rightarrow h = s_{high} - s_{low}$$

Since the underlying data behind the OHLC bars is characterized by spread and volatility as the only relevant parameters beside price, the height of the bars h must be entirely determined by them. Given as a function of time interval, it is an extension of spread, forming the **spread curve** of price uncertainty[12]:

---

[11] The square root law is the source of common mistake in risk management when at time horizon $T = 0$ securities are thought to carry no potential for losses.

[12] Not to be confused with yield spread and credit spread.

$$h = h(\Delta, s, \sigma, T) \tag{1.7}$$

Examples of spread curves are shown in FIGURE 1.7 for different time scales.

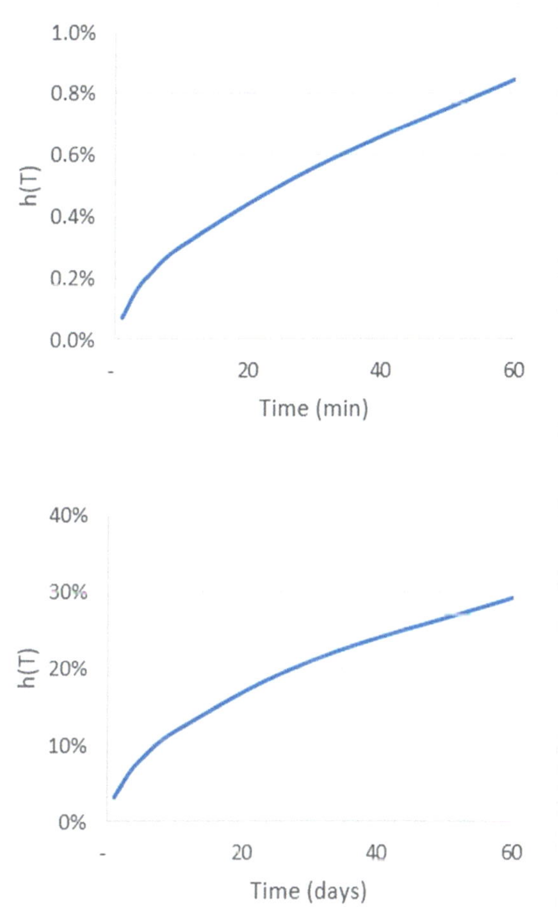

FIGURE 1.7 *Spread curves for ticker LULU for different time scales.*

If volatility properly described price uncertainty, the spread curve could be composed with it based on just one point of data, which is shown in FIGURE 1.8 with the example of spread curve for LULU stock. If a 1-minute high-low bar were scaled forward to 60 minutes in a square root law, the result would be 0.55%, which is about 35% off the real value. This difference in functional behavior is a result of the quantum nature of price formation, which prohibits price uncertainty from being lower than spread.

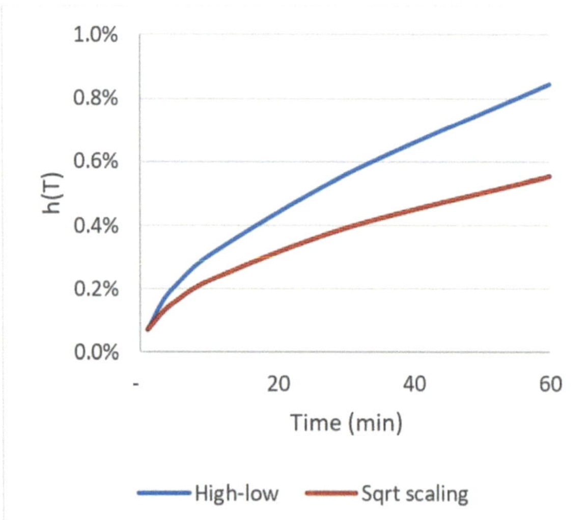

FIGURE 1.8 *Comparison between the real spread curve and classical spread curve, obtained with the square-root of time scaling.*

1.5 Uncertainty scaling

Let us see what we know about the behavior of the spread curve. At time scale smaller or comparable with the time between trades τ price uncertainty is nearly constant and equal to bid-ask spread:

$$h(\Delta, s, \sigma, T \sim \tau) \approx \Delta \tag{1.8}$$

This is the quantum regime, which was discussed earlier in this section.

In the opposite limit – time scale much larger than the time between trades – spread becomes irrelevant, its effect disappears, so only volatility and time interval remain. Like volatility, the high-low bars scale as a square root of time:

$$h(\Delta, s, \sigma, T \gg \tau) \sim s\sigma\sqrt{T} \tag{1.9}$$

This is the stochastic limit, which in this book is also referred to as the diffusion limit.

Putting the two limits together, it becomes apparent that the shape of function $h(\Delta, s, \sigma, T)$ must be similar to the one shown in FIGURE 1.9.

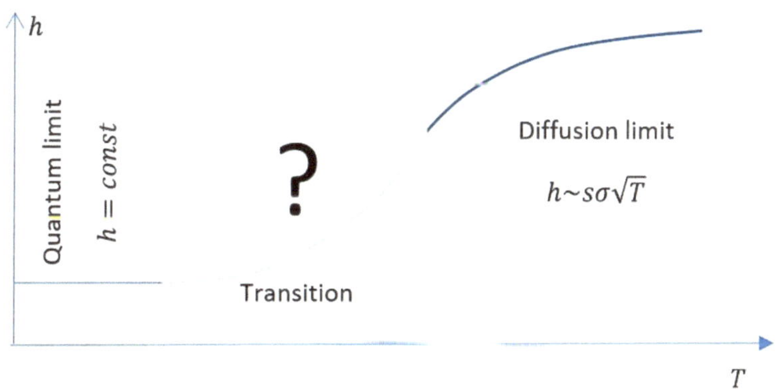

FIGURE 1.9 *Scaling of price uncertainty with time.*

Functional form of h in the transition region can only be determined by a model that encompasses the two limits, complies with them as well as the market data. Modeling the spread curve and being able to calibrate it to market data on all scales is a major point in adequate pricing, execution, and risk management.

1.6 Basic relation between spread, volatility, and volume

1.6.1 Spread as price uncertainty

Even though the square-root behavior of bar height is lost in the transition region, it is still within the order of magnitude accuracy from the actual function. This makes it possible to estimate the spread from macro parameters. Combining Eqs. (1.8) and (1.9), we get:

$$h(\Delta, \sigma, T \sim \tau) \approx \lambda s \sigma \sqrt{\tau}$$

Coefficient λ is dimensionless and has order of magnitude 1. It may vary slowly with market conditions but will maintain its order of magnitude since major dependencies of spread have already been factored out.

We can estimate τ from trading volume V and average trade size m as

$$\tau = \frac{m}{V}$$

It is important that σ and V be related to the same time interval, so if σ is daily volatility, then V has to be daily volume. Combining the two equations together, we have:

$$\Delta \approx \lambda s\sigma\sqrt{\frac{m}{V}} \tag{1.10}$$

1.6.2 Spread as straddle premium

Market making strategies that quote securities on both sides of the order book are looking to execute buy and sell trades within a narrow range. These strategies are essentially variations of bets on two conditions: security price (a) being contained in an interval within the spread (b) within the average transaction time τ. Theoretically, such bets could also be placed with options expiring after time τ. Straddle strategy is an example of such bet[13]. If security price remains within the spread, then the received premium should be sufficient to cover payables at the exercise. Therefore, the range of positive P/L should be equal to spread. This comparison is not exact and depends on the specifics of market maker's execution (Sarkissian, 2016). Exact replication may involve exotic options and more than just two-legged strategy, let alone its practical realization. Let us estimate the spread from these considerations.

Range of positive P/L for a straddle is:

$$\Delta = 2(c + p)$$

where c and p are the call and put premia received. One can estimate these premia using the approximation for ATM Black-Scholes options:

$$c_{ATM} = p_{ATM} \approx \frac{s\sigma\sqrt{\tau}}{\sqrt{2\pi}}$$

---

[13] Practical realization of such strategy is of course impossible, but this comparison serves as a good reference point for estimates.

Combining all equations gives the final result:

$$\Delta \approx \sqrt{\frac{8}{\pi}}\, s\sigma\sqrt{\tau}$$

As we can see, this method produces the same result as Eq. (1.10) with implied value of $\lambda = \sqrt{\frac{8}{\pi}} \approx 1.6$.

1.6.3 Dimensional considerations

The same relationship can be obtained from dimensional considerations. Indeed, the only parameters that could define spread are volatility σ with dimension of $\left[\frac{1}{\sqrt{time}}\right]$, trading volume V with dimension of $\left[\frac{quantity}{time}\right]$, and average executing order size m with dimension of $[quantity]$. The only combination in which they compose a spread with dimension of $[money]$ is

$$\Delta \sim s\left(\frac{m\sigma^2}{V}\right)^{\alpha}$$

where α is some power exponent. Realizing that under normal trading conditions spread must be proportional to volatility[14], we figure $\alpha = \frac{1}{2}$, and therefore:

---

[14] In some mathematical models volatility appears in various power combinations including fractional. In physics we realize that dependence on volatility may be only in form of its square. At times we can have a square of a variable depend on σ^2, which reduces to linear proportionality between the variable and σ. This is the case with spread. Dependence on volatility in powers other than square is not physical and may be encountered only in a fictional abstract world.

$$\Delta = \lambda s \sigma \sqrt{\frac{m}{V}}$$

which is the same as Eq. (1.10).

The fact that other powers of α are still permissible indicates that the corresponding factors may arise in different situations. Indeed, in market crashes, rebounds, and price corrections the dominant relationship is $\Delta \sim V$, corresponding to $\alpha = -1$[15].

Spread, volatility, and trading volume are not independent variables and are connected through Eq. (1.10). It is important that this relationship relies on no assumptions about the security, company, or market structure.

1.6.4 Relation to market data

Traders can set their own operating value of coefficient λ. Smaller λ will ensure more likely execution and larger λ will reduce execution probability. One way to find a reasonable value for coefficient λ is from market data. Traders can then calibrate their λ relative to the market.

FIGURE 1.10 demonstrates Eq. (1.10) in action. It shows the scatterplot of Δ with $\lambda = 3.5$ vs. average bid-ask spread for a number of stocks on March 16, 2016. That value of λ provides the best fit with market data.

FIGURE 1.11 shows how Eq. (1.10) works for spread dynamics. It displays calibrated Δ and average bid-ask spread in the period March 1-16, 2016 for AMZN (Amazon.com, Inc.).

---

[15] (Sarkissian, 2016)

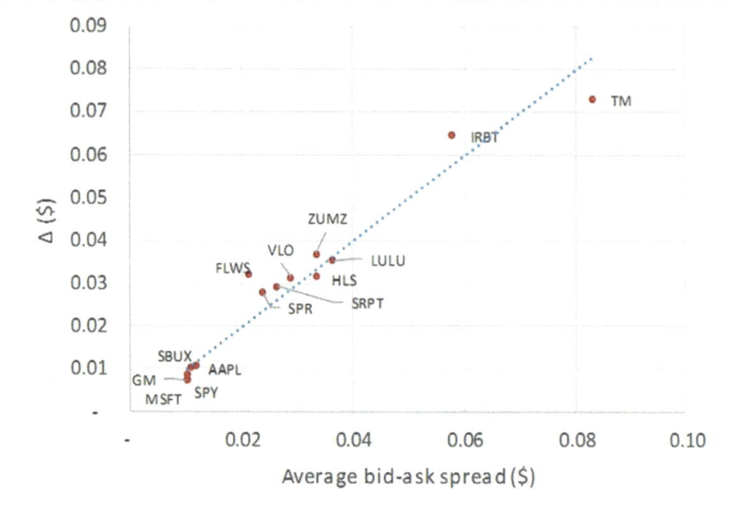

FIGURE 1.10 *Calculated spread Δ vs. average bid-ask spread on March 16, 2016. Tilt corresponds to λ = 3.5.*

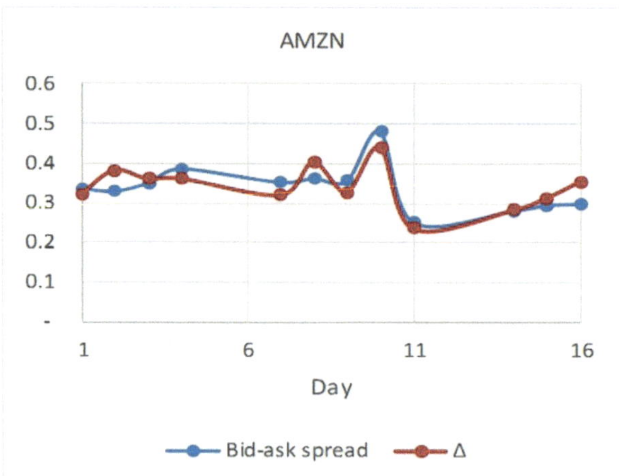

FIGURE 1.11 *Dynamics and correlation of Δ and average bid-ask spread during the period of March 1-16, 2016 for AMZN.*

1.7 Price impact with low participation

We already discussed how large trades tend to move market price in an unfavorable direction. This effect is known as price impact of trade. Consider execution of a trade with **trade quantity** Q, and let Q be positive for buy orders and negative for sell orders. **Price impact** is defined as the difference between after-trade and pre-trade prices:

$$L(Q,T) = s(Q,T) - s(0,0) \tag{1.11}$$

where $s(Q,T)$ is the price after execution and $s(0,0)$ is the price before execution. Practically, $s(0,0)$ represents price of a very small trade before execution.

A more appropriate definition of price impact is:

$$L_{eff}(Q,T) = s_{eff}(Q,T) - s(0,0) \tag{1.12}$$

based on the effective trade price at which the quantity was traded, since s_{eff} is what eventually matters in trading. For uniform execution in time $L_{eff} = \frac{L}{2}$. We are going to use the definition (1.11) as the most accepted one.

Because price movement is in the direction disadvantageous to the trader, some call this effect price slippage. To reduce impact large trades are executed in venues different from open market. One such venue is a request-for-quote (RFQ), in which trader requests quotes for his trade from other traders. Depending on the setup, quotes can be visible to the general public or hidden from it. An even more basic way involves direct non-anonymized negotiations with other traders that are known to possess or want the required block of securities.

Trade negotiations can have a wide range of outcomes and involve diverse considerations. Among them are the last trade price, recent price range, security details, optionality, covenants and guarantees, company fundamentals, credit rating, general market conjuncture, the degree of desire of parties to trade, and many more elements that may or may not be relevant.

Despite this seemingly unpredictable process, certain common practice benchmarks exist that define the negotiation window. For securities that have a shallow market the following logic is used. If trade execution in an open market would take an estimated time T by taking up all available liquidity, then the corresponding *high – low* bar is the negotiation price range. If the **trade size** is $M = |Q|$ and daily volume is V, then the time T can be estimated as $T = M/V$. From here using the arguments of Section 1.6.1, price range can be estimated as

$$s_0 - \lambda s\sigma\sqrt{\frac{M}{V}} < s < s_0 + \lambda s\sigma\sqrt{\frac{M}{V}}$$

with s_0 being either last or mid-price, and λ is again a dimensionless multiplier with order of magnitude 1. Since the trade is negotiated as a block, coefficient λ tends to 0.5 to match the effective price. Variation $\delta s = \pm\lambda s\sigma\sqrt{\frac{M}{V}}$ is considered fair to make the price attractive enough for an immediate deal. In this case δs is the price impact of the trade.

Another common strategy of executing large trades is by splitting them into smaller *child trades* and executing them sequentially. Such strategy is known to reduce impact, so that the combined impact of child orders is smaller than impact of the parent order. Whenever order splitting comes up, we will use q for the quantity and $m = |q|$ for the size of child orders generated by parent order of size $M = |Q|$.

Let us apply the square root formula to compare the two strategies. If the trade M is split into a chain of N smaller trades, so that $M = mN$ then their combined sequential impact is:

$$L_{split} \approx \lambda \sigma N \sqrt{\frac{m}{V}}$$

This is $\sqrt{N}$ times larger than the impact of an unsplit order:

$$L_{split} = \sqrt{N}\, L_{unsplit}$$

which is in contradiction with common sense. This is an indication that the square root formula is incomplete.

The root of this contradiction can be seen from the following general consideration. Price impact is a sign-dependent quantity: it is positive for buy trades and negative for sell trades. This means that for large trades, when volatility effects can be neglected, price impact is an odd function of size:

$$L(-q) = -L(q)$$

Therefore, for large trades, the Taylor series expansion of L in q must begin with the linear term:

$$L \sim q$$

As trade size grows compared to market liquidity, the square root formula must convert into linear dependence. This is exactly the behavior observed in practice.

Price impact is one of the major subjects of interest of the price formation theory. As a microstructural phenomenon it is deeply rooted in the nature of processes responsible for price formation. A more detailed analysis of price impact will be given in the following parts.

PART II

QUANTUM COUPLED-WAVE MODEL OF PRICE DYNAMICS

2.1 Introduction to coupled-wave model

Top buyers and lowest sellers are the closest candidates for a transaction. As price forms, the majority of interaction occurs between them. That is why rather than modeling the entire book it is possible to leave just the top two levels from each side and neglect the others. This 2-level approximation constitutes the basis of the **coupled-wave model**.

Essence of the coupled-wave model is demonstrated in FIGURE 2.1. While market depth certainly influences the top levels, for execution purposes its existence is insignificant if the trader's own flow is small enough not to pierce through the top levels.

If orders in the book are spaced with **order size density**

$$\rho = s\frac{dm}{ds} \tag{2.1}$$

then the model's applicability condition can be written as

$$\frac{m}{\epsilon + \sigma} \leq \rho \tag{2.2}$$

where

$$\epsilon = \frac{\Delta}{s} \tag{2.3}$$

is the relative spread and σ is volatility related to a single period.

Condition (2.2) does not mean that coupled-wave approximation is applicable only for small orders. It is related to a single period, and if liquidity renews in the next period, the condition has to be reapplied. This is why it is proper to write it in terms of order flow:

$$\frac{v}{\epsilon + \sigma} \leq \theta \tag{2.4}$$

where $v = \frac{m}{\Delta t}$ is the executing **order flow**, $\theta = \frac{\rho}{\Delta t}$ is the **order density flow** of the market and Δt is length of a single period.

Ask size	Price	Bid size
100	28.20	
900	28.15	
100	27.95	
300	27.90	
100	27.87	
	27.83	100
	27.82	400
	27.80	100
	27.79	1200
	27.78	100
	27.75	300

Coupled-wave approximation (rows 27.87 and 27.83)

FIGURE 2.1 *Demonstration of coupled-wave approximation.*

It is important to make sure that condition (2.4) is observed before applying the model. If it is violated, the coupled-wave model can lead to wrong results. For example, if spread is 2 cents, one should not calibrate the model for a flow density $\theta = 100\ shares/min/cent$ and then use it to execute an order at $v = 300\ shares/min$ flow rate.

To describe transfer between more than 2 levels one has to apply a framework that involves a larger number of levels. We will do this in PART 3. But as we will see, coupled-wave model's applicability conditions are actually wide enough to make it capable of addressing most situations.

One way to incorporate the effect of market depth is through model parameters. Dependence of model parameters on market depth can affect the size of the spread and shift execution probabilities. This in turn will have effect on price dynamics, which would be different if these effects were not included. This approach is demonstrated in Section 2.13.

Another approach to account for depth is to combine the top N individual levels into multilayer levels with effective bid and ask prices such that for them the applicability condition holds:

$$ask_{eff} = \frac{\sum_{i=1}^{N} s_{ask,i}\, m_{ask,i}}{\sum_{i=1}^{N} m_{ask,i}}$$

$$bid_{eff} = \frac{\sum_{i=1}^{N} s_{bid,i}\, m_{bid,i}}{\sum_{i=1}^{N} m_{bid,i}} \tag{2.5}$$

Here $m_{ask,i}$ and $m_{bid,i}$ are order sizes on the i-th level from the top, while $s_{ask,i}$ and $s_{bid,i}$ are the corresponding prices.

It is also possible to model the more readily available high and low levels of the OHLC bar data. In such setup the "*high*" price would correspond to "*ask*" level, and the "*low*" price would correspond to "*bid*" level. Usually bars include multiple trades, and therefore the closing price does not have to be either s_{high} or s_{low}. Instead, it can be any price within the high-low interval with a probability distribution corresponding to market data.

2.2 Price operator and price equation

To address the properties of measurement in finance, we must drift away from describing securities with price. Instead, let us introduce the **price operator *S*** whose eigenvalue spectrum represents the collection of prices, that the security can attain in trade (Sarkissian, 2013)[16]:

$$\boldsymbol{S}\psi_n = s_n\psi_n \tag{2.6}$$

This is the **Price Equation**. If price operator is represented by a matrix with elements S_{mn}, then the spectrum of prices $\{s_n\}$ can be found from the following equation:

$$\det \|S_{mn} - s_n\delta_{mn}\| = 0 \tag{2.7}$$

Eigenvectors ψ_n are complex probability amplitudes forming the basis for attaining the price corresponding to n-th state. At each moment in time the security's state is described as a linear combination (superposition) of ψ_n:

$$\Psi = \sum_{n=1}^{N} c_n\psi_n \tag{2.8}$$

[16] also (Sarkissian, 2020), (Sarkissian, 2019), (Sarkissian, 2019)

When price measurement is performed, wavefunction Ψ collapses to a specific ψ_n with its eigenvalue s_n as a result of price measurement occurring with probability

$$p_n = |c_n|^2 \tag{2.9}$$

Obviously, price operator $\boldsymbol{S}$ must be Hermitian since price is a real number. To account for fluctuations, price operator has to be stochastic:

$$\boldsymbol{S}(t + \delta t) = \boldsymbol{S}(t) + \boldsymbol{\delta S}(t)$$

Price operator can be represented by an $N \times N$ matrix corresponding to N price levels. In coupled-wave model the security has only two states: one with price equal to bid, and one with price equal to ask. Accordingly, probability amplitude has two components, and price operator is a 2×2 matrix:

$$\Psi = \begin{pmatrix} \psi_+ \\ \psi_- \end{pmatrix} \quad \text{and} \quad \boldsymbol{S} = \begin{pmatrix} s_{11} & s_{12} \\ s_{12}^* & s_{22} \end{pmatrix} \tag{2.10}$$

Writing Eq. (2.6) in matrix form we have:

$$\begin{pmatrix} s_{11} & s_{12} \\ s_{12}^* & s_{22} \end{pmatrix} \begin{pmatrix} \psi_+ \\ \psi_- \end{pmatrix} = s_{ask/bid} \begin{pmatrix} \psi_+ \\ \psi_- \end{pmatrix} \tag{2.11}$$

The diagonal elements correspond to an intrinsic price split, which can exist even in the absence of trading. This split can include liquidity costs, transaction costs, holding premium, etc., or be equal to zero. The off-diagonal elements correspond to interaction between price levels (coupling) and represent the coupling strength. These elements arise from trading activity and reflect price uncertainty due to probability transfer between levels.

Eigenvalues s_{ask} and s_{bid} are then expressed through matrix elements as

$$s_{ask} = \frac{s_{11} + s_{22}}{2} + \sqrt{\left(\frac{s_{11} - s_{22}}{2}\right)^2 + |s_{12}|^2} \qquad (2.12)$$

$$s_{bid} = \frac{s_{11} + s_{22}}{2} - \sqrt{\left(\frac{s_{11} - s_{22}}{2}\right)^2 + |s_{12}|^2} \qquad (2.13)$$

These equations can be rewritten as

$$s_{ask} = s_{mid} + \frac{\Delta}{2} \quad \text{and} \quad s_{bid} = s_{mid} - \frac{\Delta}{2} \qquad (2.14)$$

to represent the two prices spaced out by the spread

$$\Delta = \sqrt{(s_{11} - s_{22})^2 + 4|s_{12}|^2} \qquad (2.15)$$

around the mid-price

$$s_{mid} = \frac{s_{11} + s_{22}}{2} = \frac{s_{bid} + s_{ask}}{2}$$

Trade price of the period is then chosen between s_{bid} and s_{ask} according to probability of their occurrence:

$$s_{trade} = \begin{bmatrix} s_{bid} & \text{with probability } |c_{bid}|^2 \\ s_{ask} & \text{with probability } |c_{ask}|^2 \end{bmatrix} \qquad (2.16)$$

This process repeats in every step while updating the matrix elements of the price operator.

Trivial solution with single price s can be produced by the following price operator:

$$\boldsymbol{S} = \begin{pmatrix} s & 0 \\ 0 & s \end{pmatrix} \qquad (2.17)$$

2.3 Return and spread operators

Price operator is a good point for introduction, but for practical use it is more convenient to use the **return operator**:

$$\boldsymbol{R} = \frac{1}{\tau}\left(\frac{\boldsymbol{S}}{s} - \boldsymbol{1}\right) \tag{2.18}$$

where s is the current price and τ is time between trades. Return operator is a generalization of simple return and is useful to describe price evolution in a single time period relative to the previously measured price. Because there is a one-to-one correspondence between price and return, we will often talk about price and return measurement interchangeably[17].

Operators in coupled-wave model relate to a single period. We can therefore take $\tau = 1$ for convenience, just as it is done with traditional returns. The general form of return operator for a 2-level system is:

$$\boldsymbol{R} = \boldsymbol{1}z + \frac{1}{2}\begin{pmatrix} \xi & \kappa \\ \kappa^* & -\xi \end{pmatrix} \tag{2.19}$$

where z, ξ and κ are parameters and $\boldsymbol{1}$ is the identity matrix.

The $\boldsymbol{1}z$ term is the diffusion term. It describes a common shift of price level. The second term is responsible for formation of spread around the new price level. It is therefore the spread term.

Parameters z, ξ and κ describe price formation at each step. The first two of them are real and κ can be complex. These parameters can incorporate order book data, order flow data, interaction between orders, etc., and can obey separate, the so-called “material

---

[17] Referring to it in the quantum-mechanical sense, because price and return operators commute.

equations". If no particular factors are important, z, ξ and κ can be taken simply as normally distributed random numbers with zero means:

$$\xi \sim N(0, \sigma_\xi),\ \kappa \sim CN(0, \sigma_\kappa) \text{ and } z \sim N(0, \sigma_z)$$

Their values update at every step of price evolution.

Often, we want to ignore common shift and focus on the spread term. For that purpose, we turn to the **spread operator**

$$\boldsymbol{E} = \frac{1}{2}\begin{pmatrix} \xi & \kappa \\ \kappa^* & -\xi \end{pmatrix} \tag{2.20}$$

Its eigenvalues are $\pm\frac{\epsilon}{2}$ counting half spread to each side, where the relative spread is equal

$$\epsilon = \sqrt{\xi^2 + |\kappa|^2} \tag{2.21}$$

Normalized eigenvectors of the spread operator are:

$\psi_{ask} = \frac{1}{\sqrt{2\epsilon(\epsilon+\xi)}}\begin{pmatrix} \epsilon + \xi \\ \kappa^* \end{pmatrix}$ with eigenvalue $+\epsilon$ and

$\psi_{bid} = \frac{1}{\sqrt{2\epsilon(\epsilon+\xi)}}\begin{pmatrix} -\kappa \\ \epsilon + \xi \end{pmatrix}$ with eigenvalue $-\epsilon$

If security is in a superposition state

$$\Psi = c_{ask}\psi_{ask} + c_{bid}\psi_{bid} \tag{2.22}$$

the expectation values of spread and return operators are:

$$\langle\Psi|\boldsymbol{E}|\Psi\rangle = \frac{\epsilon}{2}\left(|c_{ask}|^2 - |c_{bid}|^2\right) = \frac{\epsilon}{2} I \tag{2.23}$$

$$\langle\Psi|\boldsymbol{R}|\Psi\rangle = z + \frac{\epsilon}{2}I \tag{2.24}$$

Here quantity $I = |c_{ask}|^2 - |c_{bid}|^2$ is the difference between up and down execution probabilities and is called **execution imbalance**. This quantity plays an important role in price formation and will come up further on numerous occasions.

For a security with randomly changing state Ψ_r the average over all possible realizations is $\langle\langle\Psi_r|\boldsymbol{R}|\Psi_r\rangle\rangle_r = 0$ and

$$\langle\langle\Psi_r|\boldsymbol{R}^2|\Psi_r\rangle\rangle_r = \sigma_z^2 + \left(\frac{\epsilon}{2}\right)^2 \tag{2.25}$$

This shows how in coupled-wave model price can evolve without the diffusion term ($z = 0$).

2.4 Trading algebra

Return operator can be universally decomposed through Pauli matrices as

$$\boldsymbol{R} = \mathbf{1}z + \frac{1}{2}\begin{pmatrix} \xi & \kappa \\ \kappa^* & -\xi \end{pmatrix} = \mathbf{1}z + \sum_{i=1}^{3} A_i\boldsymbol{\sigma}_i \tag{2.26}$$

where the Pauli matrices are

$$\boldsymbol{\sigma}_1 = \begin{pmatrix} 0 & 1 \\ 1 & 0 \end{pmatrix} \quad \boldsymbol{\sigma}_2 = \begin{pmatrix} 0 & -i \\ i & 0 \end{pmatrix} \quad \boldsymbol{\sigma}_3 = \begin{pmatrix} 1 & 0 \\ 0 & -1 \end{pmatrix} \tag{2.27}$$

Consider the basis vectors corresponding to "*ask*" and "*bid*" eigenstates:

$$\phi_+ = \begin{pmatrix} 1 \\ 0 \end{pmatrix} \quad \text{and} \quad \phi_- = \begin{pmatrix} 0 \\ 1 \end{pmatrix} \tag{2.28}$$

Matrix $\boldsymbol{\sigma_3}$ does not change the state probabilities, because:

$$\boldsymbol{\sigma_3}\phi_+ = \phi_+ \text{ and } \boldsymbol{\sigma_3}\phi_- = -\phi_- \tag{2.29}$$

It therefore acts as an elementary spread operator, setting the upper and lower states apart. Diagonalized spread operator can be written as

$$\boldsymbol{E_0} = \frac{\epsilon}{2}\boldsymbol{\sigma_3} \tag{2.30}$$

Matrix $\boldsymbol{\sigma_2}$ transforms the vectors into each other:

$$\boldsymbol{\sigma_2}\phi_+ = i\phi_- \text{and } \boldsymbol{\sigma_2}\phi_- = -i\phi_+ \tag{2.31}$$

so that a state with price s_{ask} transforms into a state with price s_{bid} and vice versa. It therefore acts as the **trade operator** executing 1 unit of security in a given direction. Impact of trading quantity q is described by the **trade quantity operator** $\boldsymbol{Q} = q\boldsymbol{\sigma_2}$:

$$\boldsymbol{Q} = \begin{pmatrix} 0 & -iq \\ iq & 0 \end{pmatrix} \tag{2.32}$$

Then the traded quantity with direction appears as a result of its action:

$$\boldsymbol{Q}\phi_+ = iq\phi_- \text{ and } \boldsymbol{Q}\phi_- = -iq\phi_+ \tag{2.33}$$

Eigenvectors of operator $\boldsymbol{Q}$ are:

$$\chi_+ = \frac{1}{\sqrt{2}}\begin{pmatrix} 1 \\ i \end{pmatrix} \text{ and } \chi_- = \frac{1}{\sqrt{2}}\begin{pmatrix} 1 \\ -i \end{pmatrix} \tag{2.34}$$

and eigenvalues are $+q$ and $-q$ respectively. These eigenvectors written in the basis of $\phi_{\pm}$ are just equally weighted linear combinations:

$$\chi_{+} = \frac{1}{\sqrt{2}}(\phi_{+} + i\phi_{-}) \tag{2.35}$$

$$\chi_{-} = \frac{1}{\sqrt{2}}(\phi_{+} - i\phi_{-}) \tag{2.36}$$

At the time of trade the security is in a superposition state with equally represented "*bid*" and "*ask*" states. How does this go along with observation of PART 1 that "the only time when it is valid to safe that security has a definite price is at the time of trade"? This seeming paradox has the following solution. Trade as an act of price measurement does produce a single number. However, as we have also seen in PART 1, repeated measurements generally produce different results. Even though price of a certain trade is definite, it is not a guarantee that price of another trade will be the same. Equations (2.35) and (2.36) express that fact. In a more specific way, this statement can be expressed as the uncertainty principle between accuracy of price and traded quantity.

Matrix $\boldsymbol{\sigma_1}$ acts similarly to $\boldsymbol{\sigma_2}$. Like the trade operator it switches ϕ_{+} and ϕ_{-}. In this framework it is redundant as it can be expressed through the commutator of $\boldsymbol{\sigma_2}$ and $\boldsymbol{\sigma_3}$. The return operator can be written using only $\boldsymbol{\sigma_2}$ and $\boldsymbol{\sigma_3}$ as

$$\boldsymbol{R} = \mathbf{1}z + \frac{\xi}{2}\boldsymbol{\sigma_3} - \frac{\kappa''}{2}\boldsymbol{\sigma_2} - i\frac{\kappa'}{4}[\boldsymbol{\sigma_2}\boldsymbol{\sigma_3}] \tag{2.37}$$

where κ' and κ'' are real and imaginary parts of κ: $\kappa = \kappa' + i\kappa''$.

2.5 Uncertainty relation

Operators $\boldsymbol{Q}$ and $\boldsymbol{E_0}$ do not commute:

$$[\boldsymbol{Q}\boldsymbol{E_0}] = \boldsymbol{Q}\boldsymbol{E_0} - \boldsymbol{E_0}\boldsymbol{Q} = i\epsilon q\, \boldsymbol{\sigma_1} \tag{2.38}$$

which means that the corresponding quantities cannot be measured simultaneously. Yet, these quantities do not produce an uncertainty relation, since their standard deviations are finite. Security can be brought to a clean state with respect to one of the operators, while the other quantity's variance remains finite since there are only two levels. Therefore, the lowest limit for mutual product of standard deviations is zero:

$$\sigma_\epsilon \sigma_q \geq 0 \tag{2.39}$$

Here σ_ϵ and σ_q are the relative standard deviations of spread and trade quantity. This relationship can be formally found by considering a superposition state:

$$\Psi = a_+\phi_+ + a_-\phi_- \tag{2.40}$$

Amplitudes a_+ and a_- can be represented as

$$a_+ = cos\left(\frac{\alpha}{2}\right)e^{\frac{i\beta}{2}} \quad \text{and} \quad a_- = sin\left(\frac{\alpha}{2}\right)e^{-\frac{i\beta}{2}} \tag{2.41}$$

where α is a parametrizing angle and β is the mutual phase difference between states. Then we must have

$$\sigma_\epsilon = \frac{\epsilon}{2} sin(\alpha) \text{ and } \sigma_q^2 = q^2[1 - sin^2(\alpha)\, sin^2(\beta)] \tag{2.42}$$

Minimum σ_q is achieved when $\beta = \pm\frac{\pi}{2}$. It is easy to see that product $\sigma_\epsilon\sigma_q$ can be made zero by selecting $\alpha = 0, \pm\frac{\pi}{2}, \pm\pi$. These states correspond to eigenvectors or operators $\boldsymbol{E_0}$ and $\boldsymbol{Q}$. In superposition states the uncertainty is always bigger than the minimum threshold.

The correct uncertainty relation should be adjusted for finite ranges of corresponding quantities:

$$\frac{\sigma_\epsilon}{\sqrt{\left(\frac{\epsilon}{2}\right)^2 - \sigma_\epsilon^2}} \frac{\sigma_q}{\sqrt{q^2 - \sigma_q^2}} \geq 1 \tag{2.43}$$

Equality sign is achieved when $\beta = \pm\frac{\pi}{2}$, i.e. when phase difference between the two states is $\frac{\pi}{2}$.

2.6 Spread and its statistics

Based on their description as eigenvalues of the spread operator, spread statistics can be calibrated to market data. Same is true about the OHLC bars, based on the following correspondence:

$$s_{ask} \to s_{high}$$

$$s_{bid} \to s_{low}$$

$$\epsilon = \frac{s_{ask} - s_{bid}}{s} \to h = \frac{s_{high} - s_{low}}{s}$$

Since ξ and κ enter Eq. (2.21) symmetrically, they cannot be distinguished. It is unessential for spread statistics purposes. Difference between ξ and κ plays role in execution probabilities, which is how it can be obtained.

As an example, probability distributions of bid-ask spread for TSLA and ALNY tickers along with the calibrated model distributions are shown in FIGURE 2.2.

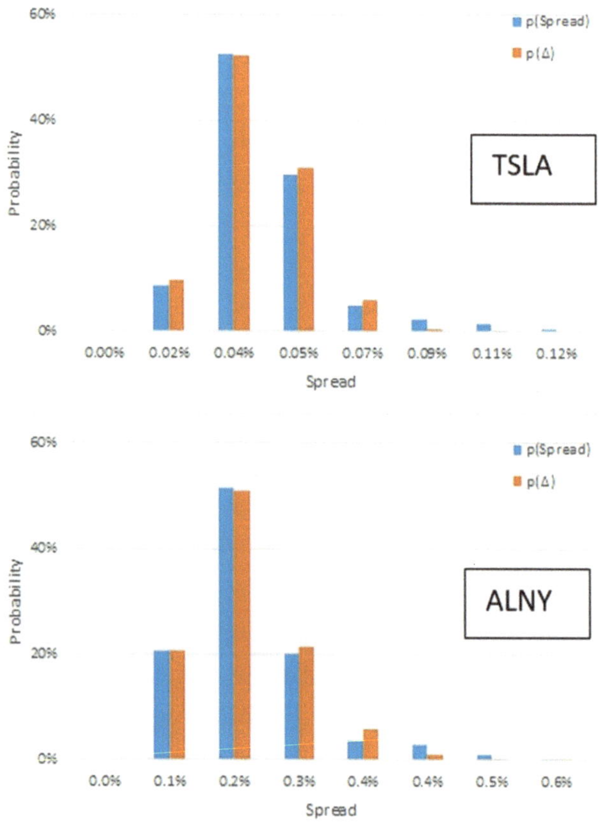

FIGURE 2.2 *Bid-ask spread probability distributions and the corresponding fits for Tesla Inc. (TSLA) and Alnylam Pharmaceuticals, Inc. (ALNY).*

In the simplest setup of coupled-wave model matrix elements of the spread operator are normally distributed random numbers. If their average values are zero, $\bar{\xi} = 0$ and $\bar{\kappa} = 0$, the probability distribution of ϵ is described by equation:

$$P(\epsilon) = \frac{1}{\sigma_\xi \sigma_\kappa} \epsilon e^{-a\epsilon^2} I_0(b\epsilon^2) \tag{2.44}$$

where

$$a = \frac{1}{4}\left(\frac{1}{\sigma_\xi^2} + \frac{1}{\sigma_\kappa^2}\right) \quad \text{and} \quad b = \frac{1}{4}\left(\frac{1}{\sigma_\xi} - \frac{1}{\sigma_\kappa^2}\right) \tag{2.45}$$

and $I_0(x)$ is the modified Bessel function of the first kind, and which has the following integral representation:

$$I_0(x) = \frac{1}{\pi} \int_0^\pi e^{x \cos(\phi)} d\phi \tag{2.46}$$

Appearance of $\epsilon\, e^{-a\epsilon^2}$ factor is typical of quantum chaotic systems described by random matrices of Gaussian orthogonal ensemble (Stöckmann, 2007). We can say that **spread behaves as a quantum chaotic quantity** (Sarkissian, 2013).

2.7 Price process

Until now, we were looking at price formation in a single period. To obtain price dynamics the period has to be repeated many times over with refreshed variables. Using the notations introduced in previous sections, price propagation from time $t - \delta t$ to time t can be expressed with the following formulas:

$$s_{mid}(t) = s_{last}(t - \delta t)\,[1 + z(t)] \tag{2.47}$$

$$\Delta(t) = s_{mid}(t)\sqrt{\xi(t)^2 + |\kappa(t)|^2} \tag{2.48}$$

$$s_{ask}(t) = s_{mid}(t) + \frac{\Delta(t)}{2} \tag{2.49}$$

$$s_{bid}(t) = s_{mid}(t) - \frac{\Delta(t)}{2} \tag{2.50}$$

Trade price is then chosen between the s_{bid} and s_{ask} according to probability of their occurrence:

$$s_{last}(t) = \begin{bmatrix} s_{bid}(t) \text{ with probability } |\psi_{bid}(t)|^2 \\ s_{ask}(t) \text{ with probability } |\psi_{ask}(t)|^2 \end{bmatrix} \tag{2.51}$$

This model includes the traditional Wiener process as a trivial case. Indeed, when $\xi = \kappa = 0$ spread equals zero, and therefore

$$s_{last}(t) = s_{last}(t - \delta t) + z(t)\, s_{last}(t - \delta t) \tag{2.52}$$

which is the traditional Wiener process for s_{last}. Therefore, the correspondence principle is observed.

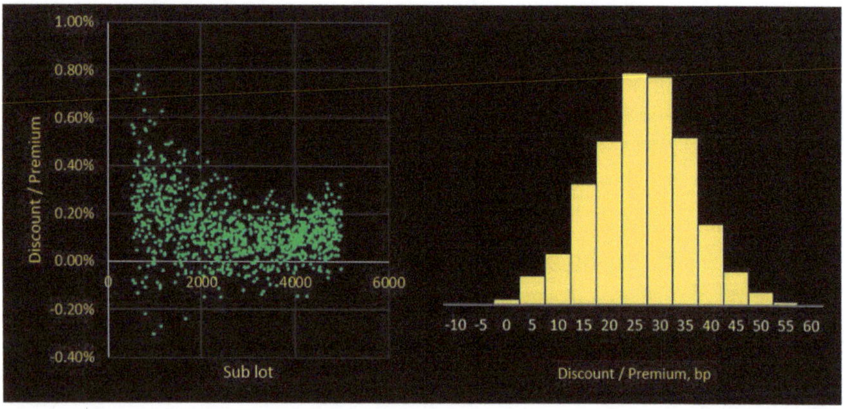

FIGURE 2.3 *Price variation and optimization with coupled-wave model.*

By running all possible price scenarios, statistics of final outcomes can be produced, such as shown in FIGURE 2.3. Depending on requirements, price dynamics can include impact of an executing order, market reaction to it, liquidity variation with time and price, self-action effects, etc.

The only missing part by now is the execution probabilities from Eq. (2.51). There are different ways to obtain them. Execution probabilities can be determined from state dynamics corresponding to the return operator. In this case market events have to be built into the return operator. This method is further developed in the current Part.

Another way to determine execution probabilities is through considering statistical ensembles of securities and studying impact of externally imposed conditions on them. This approach is developed in PART 4.

2.8 State dynamics

Price dynamics depends not only on positions of bid and ask levels, but also on selection between them. Equations (2.47) - (2.51) allow to model price dynamics if state probabilities are known. To fill the missing part, we need to know how the state of security evolves.

To be able to derive the dynamic equation for Ψ we have to rely on superposition principle. There is no reasoning allowing to deduce this principle and, in a sense, it must be postulated in finance just the way it was postulated in quantum mechanics with its validity confirmed by subsequent observations.

Strictly speaking, the set of states involved in superposition must include all possible states, particularly those associated with corporate actions, such as default, M&A, early redemption, delisting, etc. Since these are rare, they can be overlooked. Other special states are those without bidders or without sellers. These states are already included in the framework with $s_{bid} = 0$ and $s_{ask} = \infty$.

Proceeding with standard derivation procedure, we can say that the derivative $\frac{\partial \Psi}{\partial t}$ must come as a result of linear combination of the components of probability amplitude:

$$\frac{\partial \Psi}{\partial t} = \hat{T}\Psi$$

where $\hat{T}$ is some linear operator. If all states ϕ_n are taken into account as noted above, this equation must conserve total probability at all times: $p = \sum_n |\phi_n|^2 = 1$. By requiring that $\frac{\partial p}{\partial t} = 0$, it is easy to show that the equation must take form:

$$i\tau s \frac{\partial \Psi}{\partial t} = \boldsymbol{S}\Psi \tag{2.53}$$

where τ is the characteristic time with the meaning of time between trades, and operator $\boldsymbol{S}$ has to be identified with the price operator to allow recovery of stationary trivial solution with zero spread and single price:

$$\Psi(t) = e^{-i\frac{t}{\tau}}\,\Psi(0)$$

Equation (2.53) describes propagation of probability amplitude in time. Note, that its applicability is not limited by only 2 levels and can include as many levels as necessary.

Since $\boldsymbol{S}$ fluctuates, probability amplitude has to be propagated in steps δt sufficiently small that matrix elements of $\boldsymbol{S}$ remain constant. For a single step we can write the solution of Eq. (2.53) as

$$\Psi(\delta t) = e^{-i\frac{\boldsymbol{S}}{s}\frac{\delta t}{\tau}}\Psi(0) \tag{2.54}$$

For a timeframe consisting of N time steps we can write:

$$\Psi^{(k)}(t) = e^{-i\frac{\boldsymbol{S}_N^{(k)}}{s}\frac{\delta t_N}{\tau}} e^{-i\frac{\boldsymbol{S}_{N-1}^{(k)}}{s}\frac{\delta t_{N-1}}{\tau}} \dots e^{-i\frac{\boldsymbol{S}_1^{(k)}}{s}\frac{\delta t_1}{\tau}}\Psi(0) \tag{2.55}$$

where $\boldsymbol{S}_n^{(k)}$ is a particular, k-th realization of operator $\boldsymbol{S}$ at the n-th step and $\Psi^{(k)}(t)$ means the final probability amplitude resulting from that realization. Finally, to find the expected probability distribution one has to average over the ensemble of all realizations:

$$P(t) = \langle\left|\Psi^{(k)}(t)\right|^2\rangle_k \tag{2.56}$$

The elementary steps in Eq. (2.55) can be performed analytically using the Lagrange–Sylvester formula for matrix exponential:

$$\Psi(\delta t) = \left[\sum_{j=1}^{N} e^{-i\frac{s_j\delta t}{s\tau}} \prod_{\substack{n=1 \\ n\neq j}}^{N} \frac{\boldsymbol{S} - s_n\mathbf{1}}{s_j - s_n}\right]\Psi(0) \tag{2.57}$$

where the k-th realization is implied but was omitted for clarity. Also, here $\mathbf{1}$ is the identity operator and price spectrum $\{s_j\}$ is determined by equation Eq. (2.7).

2.9 Coupled-wave dynamics

Dynamics of probability amplitudes for a two-level system is governed by the following system of equations[18]:

$$i\tau s\frac{d\Psi_{ask}}{dt} = s_{11}\Psi_{ask} + s_{12}\Psi_{bid} \tag{2.58}$$

$$i\tau s\frac{d\Psi_{bid}}{dt} = s_{12}^{*}\Psi_{ask} + s_{22}\Psi_{bid} \tag{2.59}$$

These are the coupled-wave equations that couple together the components of probability amplitude. Written in terms of the return operator and after a substitution $\Psi = e^{-i\frac{t}{\tau}}\psi$, they become:

$$2i\tau\frac{d\psi_{ask}}{dt} = \xi\psi_{ask} + \kappa\psi_{bid} \tag{2.60}$$

$$2i\tau\frac{d\psi_{bid}}{dt} = \kappa^{*}\psi_{ask} - \xi\psi_{bid} \tag{2.61}$$

where τ was brought back to balance time dimensions. We also omitted the diffusion term since it can be easily eliminated by an oscillatory substitution and does not affect probability transfer.

For time interval in which matrix elements of return operator can be considered constant this system of equations has the following solution[19] expressed through model parameters ξ and κ:

$$\psi_{ask}(t) = \left[cos(\phi) - i\frac{\xi}{\epsilon}sin(\phi)\right]\psi_{ask}(0) - \\ -i\frac{\kappa}{\epsilon}sin(\phi)\psi_{bid}(0) \tag{2.62}$$

---

[18] (Sarkissian, 2013), (Sarkissian, 2020)

[19] (Zel'dovich, et al., 2004), (Zel'dovich, 2008)

$$\psi_{bid}(t) = -i\frac{\kappa^*}{\epsilon}sin(\phi)\psi_{ask}(0) + \\ + \left[cos(\phi) + i\frac{\xi}{\epsilon}sin(\phi)\right]\psi_{bid}(0) \tag{2.63}$$

where $\phi = \frac{\epsilon}{2}\frac{t}{\tau}$.

These equations have to be applied step after step, while updating the coefficients ξ and κ at each step. Knowing the execution probabilities now allows to model level selection in Eqs. (2.51). An example of path generated by the model is shown in FIGURE 2.4.

FIGURE 2.4 *Typical bid, ask and trade paths in coupled-wave model.*

Equations (2.62) and (2.63) are oscillatory in nature. If ϵ were constant, execution probabilities would have oscillated with frequency $\omega = \frac{\epsilon}{2\tau}$. However, in practice we generally do not observe oscillatory behavior in the markets. To have oscillatory behavior phase of wavefunction should continuously evolve without random jitter. Instead, it gets kicked in opposite directions by different amounts with every executing trade. The result is that random phase jitter washes away interference patterns.

Different picture is observed in critical events, such as crashes, price corrections, rallies, and rebounds. These events have lengthy periods of directional order flow as sellers sell in panic or buyers pick up bargain deals. Then each subsequent trade moves the phase in the same direction, which suppresses jitter. Unless the opposite is clear from the context, we will consider normal markets with strong phase jitter.

Now is the time to discuss implications of the superposition principle used in derivation of Eq. (2.53). While the derivation did rely on linearity in Ψ, it does not mean that the model is able to describe only linear effects. Nonlinear effects can come into play through coefficients ξ and κ. Being the components of the return operator, these coefficients depend on trading environment and market microstructure. The way they were introduced here, they were assumed random reflecting the fact that the trading environment is generally random. In situations where certain non-random features become prominent, coefficients ξ and κ must follow the material equations applicable for such situations. These equations may include combinations of Ψ and Ψ^*, thus giving rise to nonlinear effects. Nonlinear coupling models are described in Section 2.13.

Dependence of probability evolution on initial conditions in Eqs. (2.62) and (2.63) may appear unnatural. Surely, the effect of initial conditions can be thought to wash away with time due to large fluctuations. But what determines the initial conditions? How important are they? How long does their effect propagate through price evolution? The answer to that lies in the ergodic properties of Eqs. (2.60) and (2.61), which are addressed in Section 2.12 and PART 4.

2.10 Execution imbalance

Rather than dealing with probabilities p_{ask} and p_{bid} it is convenient to combine them into another physically meaningful measure – the **execution imbalance**:

$$I = |\psi_{ask}|^2 - |\psi_{bid}|^2 = p_{ask} - p_{bid} \tag{2.64}$$

Execution imbalance shows how much more likely execution is to occur at the "*ask*" level as opposed to "*bid*" level. This quantity already showed up in Section 2.3 as part of expectation value of spread operator in a superposition state:

$$\langle\psi|\boldsymbol{E}|\psi\rangle = \frac{\epsilon}{2} I$$

Execution imbalance can span from 1, when orders are certain to execute at s_{ask} to -1, when orders are certain to execute at s_{bid}, and is equal to 0 when execution probabilities are equal. Given that phase is random our interest in system's state can be limited to execution imbalance.

Using execution imbalance, the probabilities can be expressed as:

$$p_{ask} = \frac{1+I}{2} \text{ and } p_{bid} = \frac{1-I}{2} \tag{2.65}$$

Accordingly, probability amplitudes are:

$$\psi_{ask} = \sqrt{\frac{1+I}{2}} e^{i\frac{\beta}{2}} \text{ and } \psi_{bid} = \sqrt{\frac{1-I}{2}} e^{-i\frac{\beta}{2}} \tag{2.66}$$

where β is the phase difference. Combining Eqs. (2.60) and (2.61) we get the dynamic equation for imbalance:

$$\tau \frac{dI}{dt} = 2\, Im(\kappa \psi^*_{ask} \psi_{bid}) \tag{2.67}$$

If $\delta\phi$ is the phase of $\kappa\psi^*_{ask}\psi_{bid}$, then

$$\frac{dI}{dt} = \frac{|\kappa| sin(\delta\phi)}{\tau} \sqrt{1 - I^2} \tag{2.68}$$

This is a much more compact equation than Eqs. (2.60) and (2.61). It turns out that modeling imbalance can be easier than modeling probability amplitudes. It is also justified considering the fact that eventually we are interested in probabilities and not the phase.

As we can see, $I(t) = \pm 1$ are solutions of equation (2.68), corresponding to $p_a = 1$ or $p_b = 1$. Note that $I = 0$ is not generally a solution.

Finally, writing imbalance as $I = sin\,(\alpha)$, we arrive at equation for α:

$$\frac{d\alpha}{dt} = \frac{|\kappa| sin\,(\delta\phi)}{\tau} \equiv K \tag{2.69}$$

With random ξ and κ the imbalance I is irregularly distributed over interval $[-1; 1]$ as shown in FIGURE 2.5 and FIGURE 2.6. But if $\xi = 0$ and $\kappa = i\kappa''$ and the phase shift between initial "*bid*" and "*ask*" states is zero, then $\delta\phi = \frac{\pi}{2}$. Then α performs a random walk uniformly covering the $[0; 2\pi)$ interval. The corresponding distribution of imbalance $\Omega(I)$ is:

$$\Omega(I) = \frac{1}{\pi\sqrt{1 - I^2}} \tag{2.70}$$

as can be seen in FIGURE 2.7.

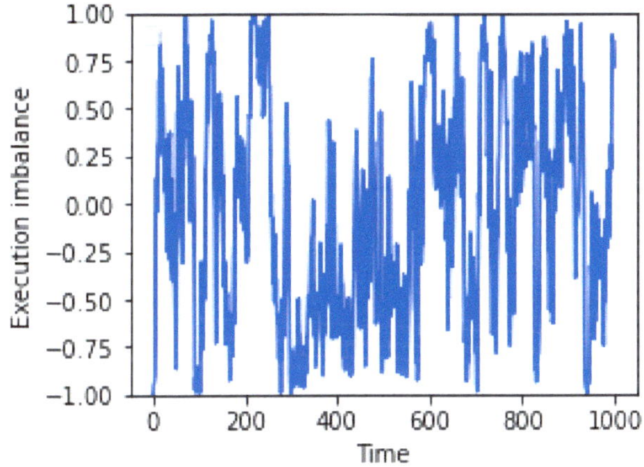

FIGURE 2.5 *Time evolution of freely evolving I with* $\xi \neq 0$.

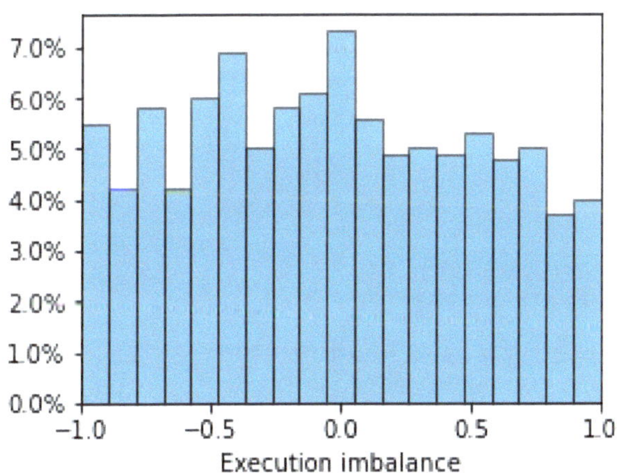

FIGURE 2.6 $\Omega(I)$ *of freely evolving I with* $\xi \neq 0$

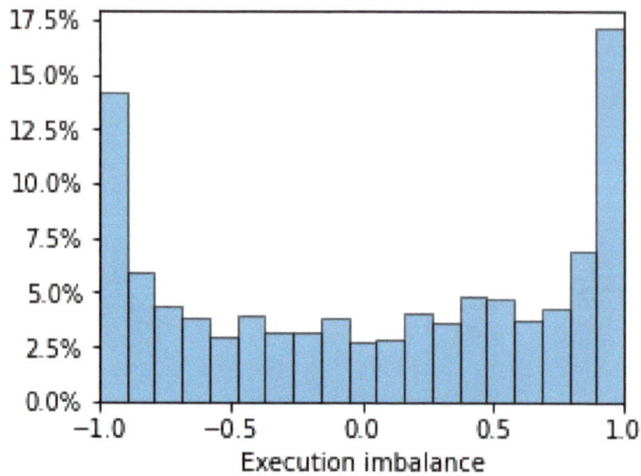

FIGURE 2.7 $\Omega(I)$ of freely *evolving* I *with* $\xi = 0$ *and imaginary* κ.

Neither of the distributions shown in FIGURE 2.6 and FIGURE 2.7 has a shape generally observed in the markets. Actual $\Omega(I)$ is localized around $I = 0$ as imbalance fluctuates around the zero point trying to drive the security back to balance. Such discrepancy points at existence of mechanisms that modify the dynamics of the system by introducing non-random elements into it (Sarkissian, 2017).

2.11 Order imbalance

The main driver in establishing market equilibrium is order imbalance, which is the difference between buy and sell sizes. Since only relative numbers matter in this process, **order imbalance** is defined as:

$$J = \frac{m_{bid} - m_{ask}}{m_{bid} + m_{ask}} \tag{2.71}$$

where m_{ask} is the selling size and m_{bid} is the buying size. Order imbalance equals $+1$ in a market without sellers ($m_{ask} = 0$), it equals -1 in a market without buyers ($m_{bid} = 0$) and equals 0 in a balanced market where sellers and buyers are trading equal quantities ($m_{ask} = m_{bid}$). In this sense the sign of order imbalance is aligned with the sign of execution imbalance.

Apparently, execution and order imbalances are related: $I = I(J)$. If the entire order imbalance is set to execute in timeframe T and $\Delta t = \frac{m_{bid}+m_{ask}}{2V}$ is the turnover time for top levels based on volume V, then

$$I \approx \frac{\Delta t}{T} J \tag{2.72}$$

Factor Δt is intended to adjust for top level thickness and accounts for the fact that out of the entire top size only $2V$ participates in execution within the time period.

The real connection between I and J is more complicated due to multiple order levels, hidden and fleeting liquidity, and collective phenomena. However, some things can be said about the relation between I and J without knowing more details.

Since flipping the book results in flipping the sign of imbalances, function $I(J)$ is an odd function: $I(-J) = -I(J)$. Therefore, for small imbalances the Taylor series expansion of $I(J)$ must begin with a linear term, just the way Eq. (2.72) already has it. Given than $|J| \leq 1$ and that the next term in the expansion is $\sim J^3$, accuracy of Eq. (2.72) is quite sufficient for most cases related to balance and liquidity.

Both I and J range in interval $[-1, 1]$. For one-sided books we must have $I(+1) = 1$ and $I(-1) = -1$. Thus, the overall shape of function $I(J)$ transpires as the one shown in FIGURE 2.8.

FIGURE 2.8 *Dependence of execution imbalance on order imbalance*

In a sense, execution imbalance is unmeasurable. Indeed, to measure execution imbalance one has to mark trades occurring in a time interval as buy or sell trades and compute their fractions. For statistical significance, the time interval has to be long enough to involve many trades. In such setup sensitivity of execution imbalance to time is lost.

To the contrary, order imbalance is observable and immediately measurable. This is why it is important to relate one to the other. We will rely on this relation further, both in terms of their values and in the sense of their probability distributions. In the latter case we will refer to $\omega(J)$ as a proxy for $\Omega(I)$.

2.12 Ergodicity and price formation

When security trades at its fair value, we must have $p_a = p_b$ and therefore $I = 0$. This does not mean that price does not move, but only that buying and selling probabilities are equal. In reality I will fluctuate around 0 being dispersed with probability distribution $\Omega(I)$. This probability distribution tends to have a more regular and narrow shape for more liquid securities than for illiquid ones. Invoking the similarity between $\Omega(I)$ and $\omega(J)$ established in the previous section, some examples are shown in FIGURE 2.9.

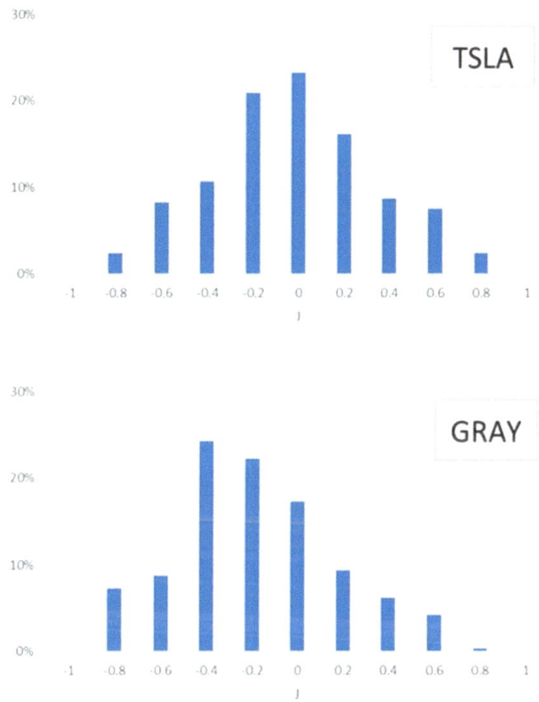

FIGURE 2.9 *$\omega(J)$ for liquid (top) and illiquid (bottom) securities*

Since $\psi(t)$ depends on initial conditions, it appears that so does $\Omega(I)$. It is certainly true for a limited observation period. As the security evolves, it goes through different microstates of its phase space, and if it is **ergodic**, it goes through all microstates (G.G. Malinetsky, 2000). Once each microstate has made its contribution to $\Omega(I)$, the initial conditions are no longer singled out, and $\Omega(I)$ must converge to a definite function, which stays invariant under transformation represented by Eqs. (2.62) and (2.63) (Sinai, 1977).

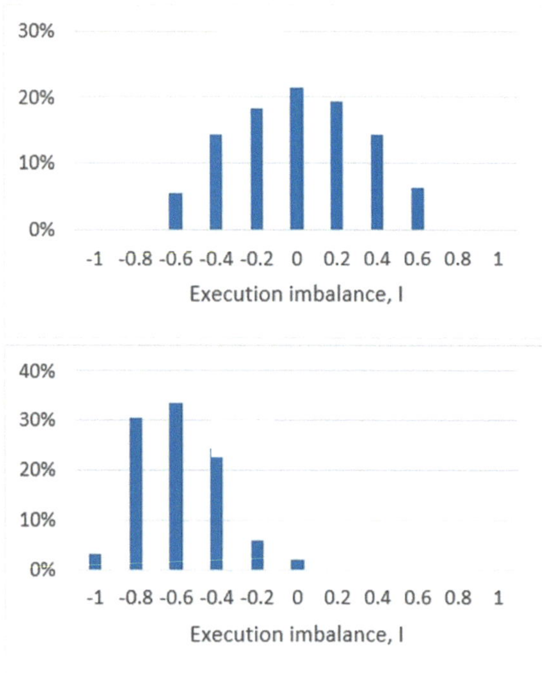

FIGURE 2.10 *Probability distribution $\Omega(I)$ in a balanced market (top), and $\Omega(I)$ in a crashing market (bottom).*

This is the generally observed picture in balanced market conditions, i.e., when the security trades around its fair value,

FIGURE 2.10 (top). A different picture arises in imbalanced situations, such as price corrections, crashes, and rallies. When the number of buyers or sellers is excessive the initial conditions are such that I is close to 1 or -1 depending on direction. Their effect in Eqs. (2.62) and (2.63) persists in time leading to an asymmetric shape of $\Omega(I)$ and enabling I to stay positive or negative for extended time. The security loses ergodicity and becomes confined to states with $I < 0$ or $I > 0$, while having the other states still available as shown in FIGURE 2.10 (b). Such ergodicity loss is accompanied by sharply rising correlations and the volatility-drawdown hysteresis, which is a typical ergodicity-breaking behavior[20].

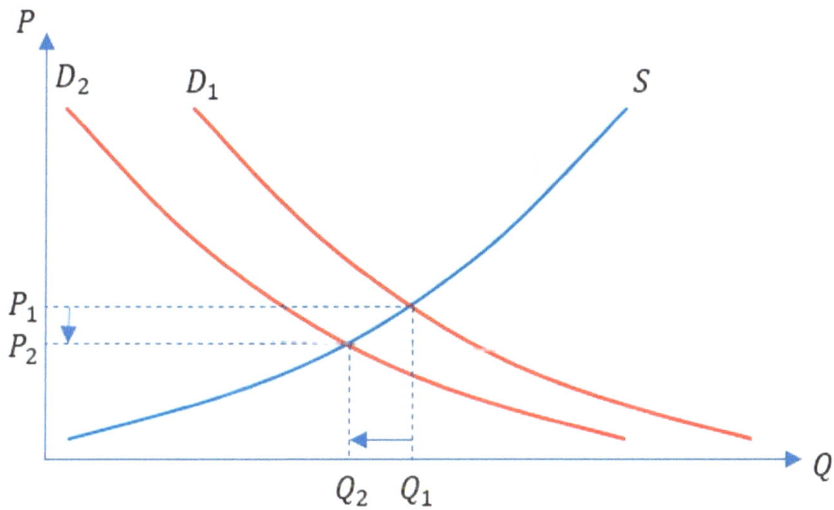

FIGURE 2.11 *Supply-Demand diagram. Source of primitive belief that when supply or demand changes the intersection point starts moving toward its new equilibrium under the influence of a mystical force.*

[20] (Sarkissian, 2019)

This consideration marks a crucial difference of quantum price formation theory with classical economic theory. Classical economics teaches that price forms as a result of balance between supply (S) and demand (D) at the intersection of supply and demand curves[21] (single number!). When either of them changes, it results in a shift of the intersection point as shown in FIGURE 2.11.

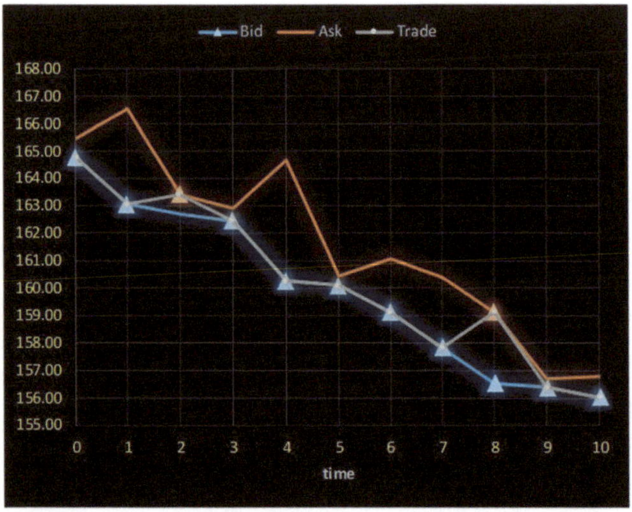

FIGURE 2.12 *Directional price move in price crash scenario. Due to ergodicity breaking most executions happen at the bid level, driving price down without application of any forces.*

This is not how it happens in quantum framework. Take a crash scenario as an example. Excessive volume of sellers shifts the security state towards the bid-level and bid price becomes the more likely outcome of measurement. Once a trade happens at that level,

[21] (J. Hirshleifer, 2005), (C. Caginalp, 2019), (C. Plott, 2017)

bid and ask orders will reposition around the new price level. Once again, probability shift makes the bid-level a more likely outcome for the next trade. Repeated over many steps, price goes down as shown in FIGURE 2.12. This happens as a result of ergodicity loss in a quantum-chaotic process without any mystical forces acting on price.

2.13 Dynamic coupling

2.13.1 Balanced coupling

In a normal market condition overbuying or overselling induces a reactive order flow that is proportionate to imbalance and acts to restore balance:

$$v = \frac{gI}{\tau} \tag{2.73}$$

Coefficient g can have any sign since it will become unimportant in the final equations. Adding this reactive flow to the random flow, represented by noise η' with standard deviation $\sigma_{\eta'}$, we can write the coupling coefficient as,

$$\kappa = gI + \eta' \tag{2.74}$$

Plugging it into Eq. (2.68), we get

$$\frac{dI}{dt} = \lambda\, I\sqrt{1 - I^2} + \eta \tag{2.75}$$

where λ is now a random number of any sign and η is a redefined noise.

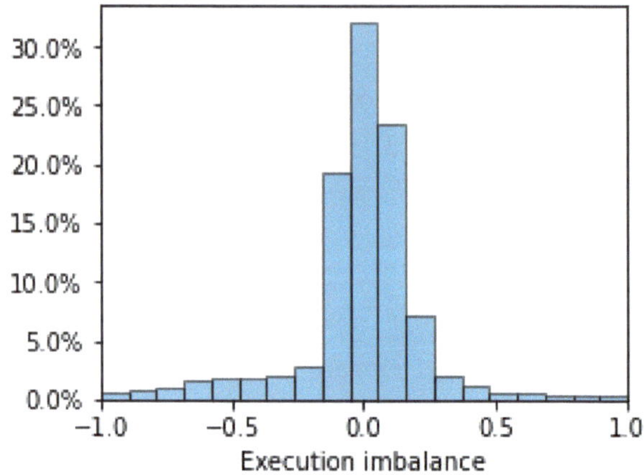

FIGURE 2.13 *Balanced* $\Omega(I)$.

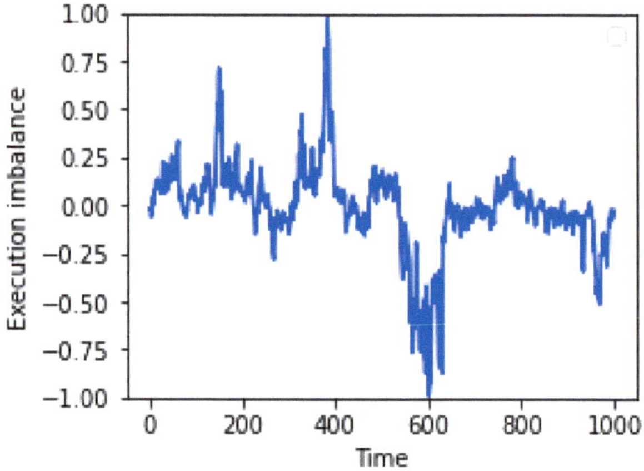

FIGURE 2.14 *Time evolution of imbalance I in balanced coupling model.*

Coupling that is proportional to I provides the stability point $I = 0$ for Eq. (2.68), which was missing with random coupling. At $I = 0$ the dynamic equations decouple, allowing each probability amplitude to propagate on its own while maintaining the balance requirement $p_a = p_b$. Such coupling produces stable $\Omega(I)$ distributions localized around $I = 0$ as is shown in FIGURE 2.13 and FIGURE 2.14.

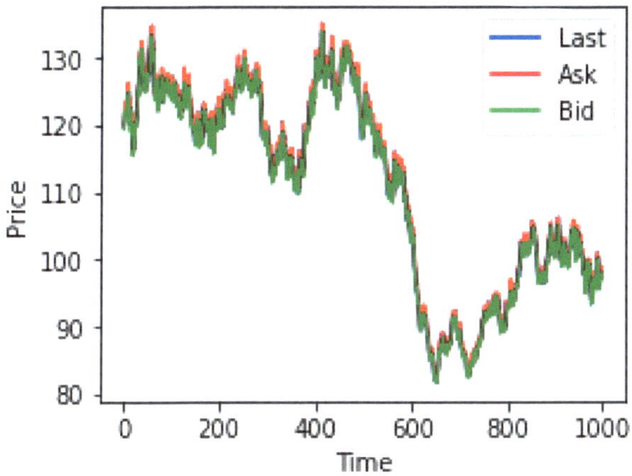

FIGURE 2.15 *Time evolution of price in balanced coupling model. Notice that the period of fall coincides with the period of negative imbalance in* FIGURE 2.14.

A few words must be said about the sign of coefficient λ. In mechanical systems mean reversion is associated with reverting force. For that purpose, reversion would clearly occur if λ was definite negative.

Quantum systems work differently. Consider for simplicity a system without noise. Time spent by the system in a state with imbalance I in a region ΔI around it, can be determined as

$$\Delta t \approx \frac{\Delta I}{|\lambda I|\sqrt{1 - I^2}} \tag{2.76}$$

From here we see that the system will spend less time in any state with $I \neq 0, \pm 1$ than in a state with $I = 0$ or $I = \pm 1$. Out of these three, the $I = \pm 1$ states are unstable with respect to fluctuations that are actually present. Getting kicked out of the imbalanced states into the state with $I = 0$ has an effect similar to reverting force, which however, happens without application of any force. This takes place with any sign of coefficient λ. An example of this effect occurring is shown in FIGURE 2.14. Corresponding price evolution is shown in FIGURE 2.15. It is noteworthy that the period of price fall coincides with the period of negative imbalance.

Reasoning behind balanced coupling makes it applicable only in normal market conditions[22], since only then market reaction induces a balancing order flow. Trader behavior during market instability is the opposite – initial imbalance induces an order avalanche that amplifies it further (Salov, 2019).

Plugging the κ from Eq. (2.74) into Eq. (2.21) we can get spread dependency on imbalance:

$$\epsilon(I) = \sqrt{\mu^2 + g^2 I^2} \tag{2.77}$$

where μ is a random number. This goes along with observations that spread increases during market crashes and rallies, which are highly imbalanced states. Yet, it does not comply with all observations.

At the turning point when crash changes into a rebound, imbalance momentarily vanishes to flip its sign. Then, according to Eq. (2.77) spread would have to exhibit a dip. Same is applicable to volatility. However, no such dip is observed. Quite the opposite, at the turning

---

[22] as opposed to critical conditions, such as in market crashes and price corrections

point volatility and spread are largest. This discrepancy does not come as a surprise since we already know that the balanced model is inapplicable in unstable markets.

2.13.2 XPM coupling

Market balance assumes no directional order flow in the absence of imbalance. That is a reasonable assumption reflecting the nature of balance. In balanced conditions coupling strength is commensurate with the price level split:

$$|\kappa| \sim |\xi|$$

During market crashes the balance at the turning point involves large order flow, so that coupling becomes prevalent:

$$|\kappa| \gg |\xi|$$

The only way that this can happen in the absence of imbalance is due to interference effects involving cross-products such as $\psi^*\psi$.

Consider the effect of just one such term on coupling:

$$\kappa = g\psi_{ask}\psi_{bid}^* \tag{2.78}$$

where g is real. With this coupling we get

$$\frac{dI}{dt} = 0$$

so that the solution is:

$$I = const$$

Coupled-wave equations under this type of coupling read:

$$2i\tau\frac{d\psi_{ask}}{dt} = (\xi + g|\psi_{bid}|^2)\psi_{ask} \tag{2.79}$$

$$2i\tau\frac{d\psi_{bid}}{dt} = (-\xi + g|\psi_{ask}|^2)\psi_{bid} \tag{2.80}$$

Their solution is:

$$\psi_{ask}(t) = e^{-\frac{i}{2}(\xi+g|\psi_{bid}|^2)\frac{t}{\tau}}\psi_{ask}(0) \tag{2.81}$$

$$\psi_{bid}(t) = e^{-\frac{i}{2}(-\xi+g|\psi_{ask}|^2)\frac{t}{\tau}}\psi_{bid}(0) \tag{2.82}$$

As we can see, only probabilities are decoupled, while the probability amplitudes are still coupled through cross-phase modulation (**XPM**), in which phases of the states are mutually affected by cross-probabilities. This coupling regime allows imbalanced states to exist for a lengthy period of time.

Spread in XPM coupling is equal:

$$\epsilon = \sqrt{\xi^2 + \frac{g^2}{4}(1 - I^2)} \tag{2.83}$$

Even though spread depends on imbalance, it remains unchanged throughout the process, since $I = const$.

2.13.3 Avalanche coupling regime

Probability decoupling occurs as long as g is real. Consider now an imaginary $g = ig''$:

$$\kappa = ig''\psi_{ask}\psi_{bid}^* \tag{2.84}$$

Then one of the probabilities amplifies and the other one attenuates depending on the sign of g'':

$$p_{ask}(t) = e^{g'' p_{bid}(t) \frac{t}{\tau}} p_{ask}(0) \tag{2.85}$$

$$p_{bid}(t) = e^{-g'' p_{ask}(t) \frac{t}{\tau}} p_{bid}(0) \tag{2.86}$$

The dynamic equation for imbalance is:

$$\tau \frac{dI}{dt} = \frac{g''}{2} \ (1 - I^2) \tag{2.87}$$

and its solution is:

$$I(t) = tanh\left(\frac{g''}{2\tau}(t - t_0)\right) \tag{2.88}$$

This process describes an avalanche order flow from one state to the other as that other state becomes populated. Remember that coupling is proportional to the order size $\kappa{\sim}iq$ and therefore $q{\sim}\psi_{ask}\psi^*_{bid}$. Thus, q is nonzero when both bid and ask levels are populated and is maximized when they are populated equally.

FIGURE 2.16 and FIGURE 2.17 show how starting with a negatively imbalanced state in which price keeps dropping, imbalance diminishes as traders begin submitting more and more buying orders ultimately flipping the imbalance and changing price direction to positive.

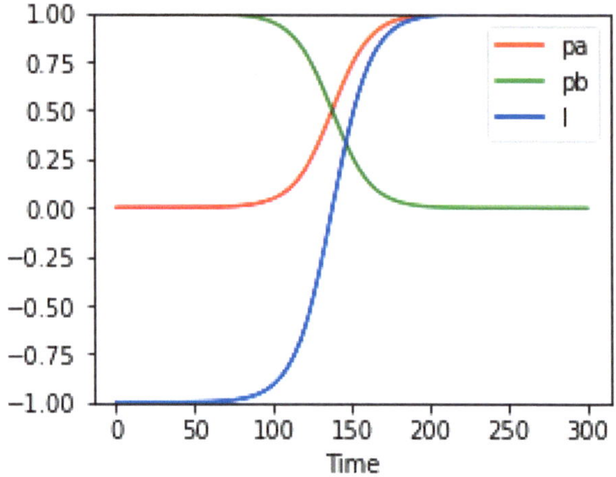

FIGURE 2.16 *Evolution of imbalance and probabilities under the avalanche coupling with* $g'' > 0$.

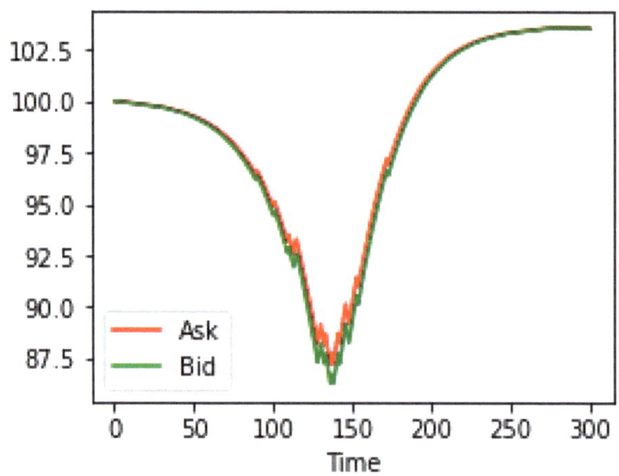

FIGURE 2.17 *Bid and ask price evolution under avalanche coupling. Price bottoms as soon as imbalance turns positive.*

If g is a complex number, its real part does not affect probability transfer. However, it does affect the spread and therefore price evolution. Writing the coupling coefficient as

$$\kappa = (g' + ig'')\psi_{ask}\psi_{bid}^{*} \tag{2.89}$$

the spread can be calculated to equal:

$$\epsilon = \sqrt{\xi^2 + \frac{|g|^2}{4}(1 - I^2)} \tag{2.90}$$

This behavior is opposite to balanced coupling in that the maximum of spread is achieved when $I = 0$.

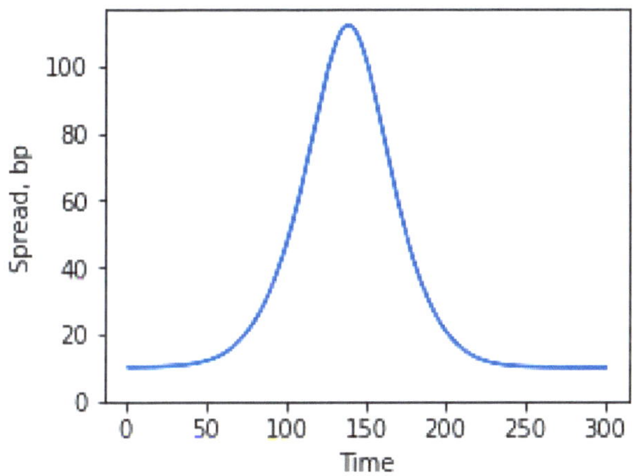

FIGURE 2.18 *Spread evolution under avalanche coupling. As security goes through the balanced state at price bottom, spread goes through the maximum.*

2.13.4 Depleting avalanche

For $g'' > 0$ order flow defined by Eq. (2.89) transfers probability from ψ_{bid} to ψ_{ask}. In financial markets such coupling cannot continue indefinitely: it can persist only to extent that the ψ_{bid} state is populated. Once the bid level is depleted, there is nothing to transfer. To reflect that fact, Eq. (2.89) must be adjusted by including the population of the bid state $|\psi_{bid}|^2$:

$$\kappa = (g' + ig'')|\psi_{bid}|^2\, \psi_{ask}\psi_{bid}^* \tag{2.91}$$

The dynamic equation now reads:

$$\tau\frac{dI}{dt} = \frac{g''}{4}\,(1 - I^2)(1 - I) \tag{2.92}$$

From here execution imbalance I obeys the following equation:

$$\frac{1 + I}{1 - I} e^{\frac{2}{1-I}} = e^{\frac{g''}{\tau}(t - t_0)} \tag{2.93}$$

Weight $|\psi_{bid}|^2$ in Eq. (2.91) has an effect of a slowing down probability transfer, which can be noticed in FIGURE 2.19. As a result, price does not restore to its previous level, FIGURE 2.20.

With this coupling spread is equal

$$\epsilon = \sqrt{\xi^2 + \frac{|g|^2}{16}(1 - I^2)(1 - I)^2} \tag{2.94}$$

Spread dynamics preserves its qualities from plain avalanche coupling, but in addition becomes asymmetric in time. Its decline in price recovery is slower than rise in a price crash. This goes along with observations and practical experience. It is out of interest here to cite academic works that confirm it.

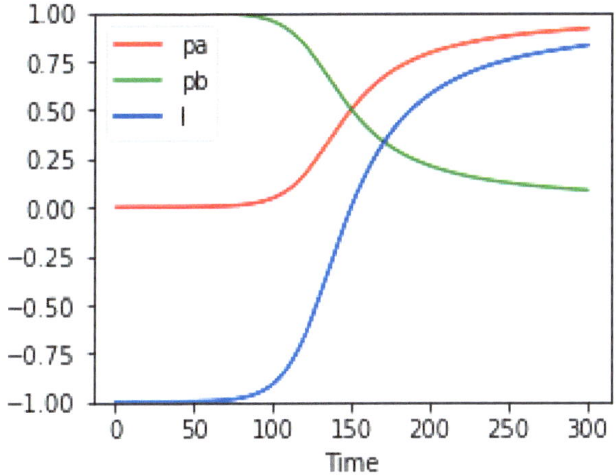

FIGURE 2.19 *Dynamics of imbalance and probabilities under the depleting avalanche coupling with g'' > 0. Transfer slows down as* p_{bid} *depletes.*

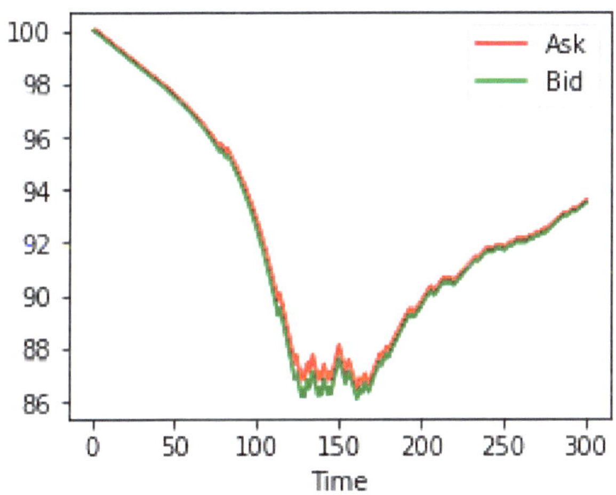

FIGURE 2.20 *Bid and ask price dynamics under depleting avalanche coupling. Price recovery lessens compared to plain avalanche regime.*

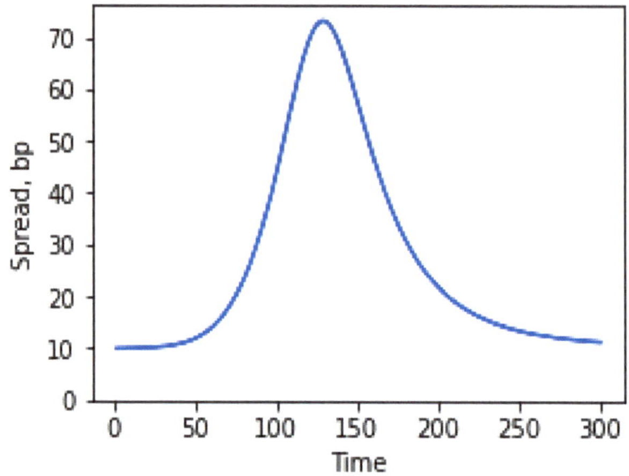

FIGURE 2.21 *Spread dynamics under depleting avalanche coupling. Spread peaks at the market bottom and its decline is slower than rise. These features go along with observations.*

2.13.5 Market reaction: conjugate order flow

In the depleting avalanche regime probability transfer from bid level to ask slows down. Yet, it goes all the way from ψ_{bid} to ψ_{ask} without leading to a stable balance. It also cannot reproduce the oscillatory pattern often observed as the balance is being established.

To describe balancing and price stabilization the opposite order flow must be taken into account. That flow is conjugate[23] to original flow and builds up as probability is transferred into *ask* state: $g^*|\psi_{ask}|^2\,\psi_{ask}^*\psi_{bid}$. Combining the two flows together, we have the coupling coefficient as:

---

[23] (Zel'dovich, et al., 2013)

$$\kappa = g|\psi_{bid}|^2 \, \psi_{ask}\psi_{bid}^* + g^*|\psi_{ask}|^2 \, \psi_{ask}^*\psi_{bid} \tag{2.95}$$

This equation unifies the above-described models in that it contains all of them as limited cases. Indeed, using notations of Eq. (2.66) and $g = |g|e^{i\phi}$, we can express the coupling coefficient as

$$\kappa = \frac{|g|}{2}\sqrt{1 - I^2}\big(cos(\phi + \beta) - iIsin(\phi + \beta)\big) \tag{2.96}$$

For small imbalance when $|I| \ll 1$, it turns into:

$$\kappa = \frac{|g|}{2}\big(cos(\phi + \beta) - iIsin(\phi + \beta)\big) \tag{2.97}$$

Here only the imaginary part depends on execution imbalance, which is sufficient, since only that part has effect on probability transfer.

Further, for $I \approx -1$ the coupling coefficient reduces to the depleting avalanche format:

$$\kappa = g|\psi_{bid}|^2 \, \psi_{ask}\psi_{bid}^* \tag{2.98}$$

and for $I = 1$ to the corresponding conjugate flow:

$$\kappa = g^*|\psi_{ask}|^2 \, \psi_{ask}^*\psi_{bid} \tag{2.99}$$

If security dynamics starts with a strong selloff as in FIGURE 2.22 and FIGURE 2.23, the selling gradually slows as sellers deplete. Simultaneously a buy flow is induced as p_{ask} grows. If the initial selling flow was strong enough, it may result in an inverted state with positive imbalance rather than just going into the balance. After an oscillation, the two flows come to balance.

Spread is linked to imbalance through the following equation:

$$\epsilon = \sqrt{\xi^2 + \frac{|g|^2}{4}(1 - I^2)[cos^2(\phi + \beta) + I^2\, sin^2(\phi + \beta)]} \qquad (2.100)$$

Its behavior corresponding to FIGURE 2.22 and FIGURE 2.23 is shown in FIGURE 2.24. The quick rise and slow decline are typical of price crashes.

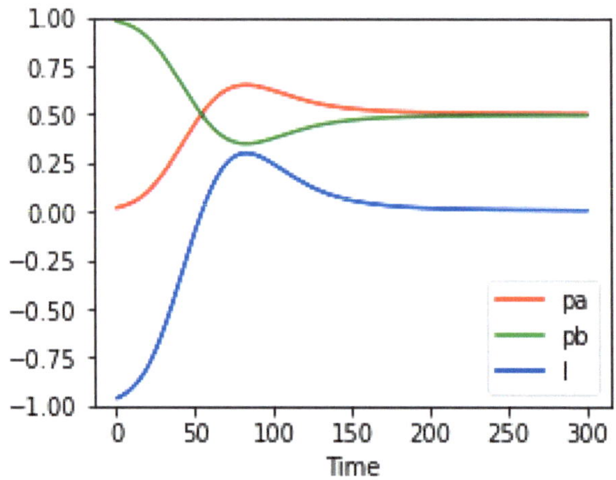

FIGURE 2.22 *Imbalance and probabilities under conjugate coupling with* $g'' > 0$ *and* $I(0) = -1$. *Imbalance goes through an oscillation before coming to zero.*

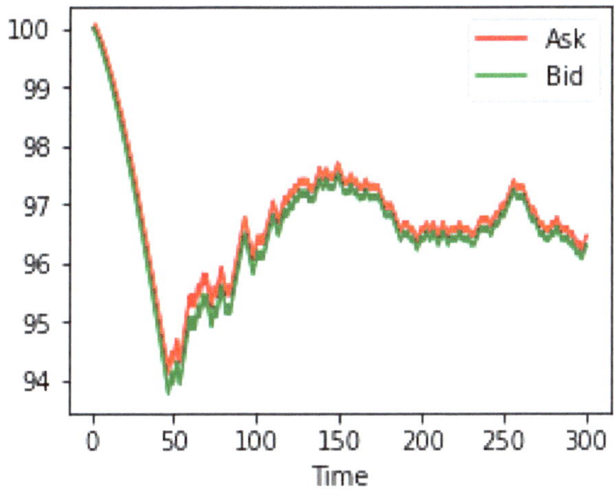

FIGURE 2.23 *Bid and ask price dynamics under conjugate coupling.*

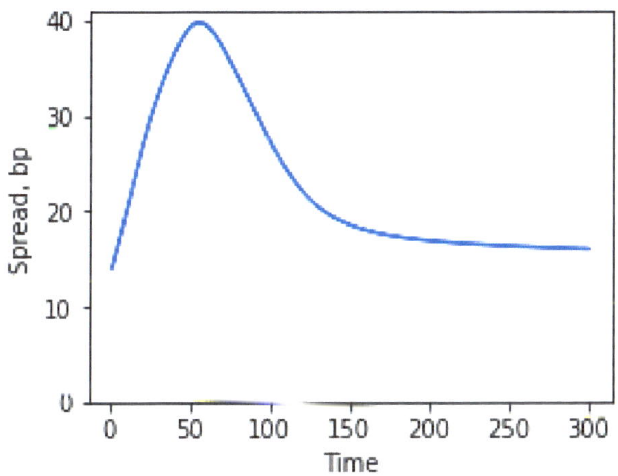

FIGURE 2.24 *Spread dynamics under conjugate coupling. Spread peaks at the market bottom and its decline is slower than rise. These features go along with observations.*

2.14 Multiperiod scaling

Return operator from Eq. (2.19) is defined by a limited set of parameters and relates to a single time interval. Evolution over multiple intervals will be described by a different return operator which, must be expressible through the same set of parameters and number of intervals. Question arises, how does a return operator scale over time?

To determine that, we must propagate the initial state over multiple periods δt:

$$\psi_T = e^{-i\boldsymbol{R}_T\delta t} \dots e^{-i\boldsymbol{R}_1\delta t}\,\psi_0 \tag{2.101}$$

where $\boldsymbol{R}_1$, … , and $\boldsymbol{R}_T$ are return operators for time periods from 1 to T. The same equation can be rewritten with a single-step propagation operator $\boldsymbol{R}$:

$$\psi_T = e^{-i\boldsymbol{R}T}\psi_0 \tag{2.102}$$

Therefore, we must have

$$e^{-i\boldsymbol{R}T} = e^{-i\boldsymbol{R}_T\delta t} \dots e^{-i\boldsymbol{R}_1\delta t} \tag{2.103}$$

which is the basis for finding $\boldsymbol{R}$. To perform multiplication, the Baker–Campbell–Hausdorff formula for matrix exponentials should be used:

$$e^{\boldsymbol{X}}\,e^{\boldsymbol{Y}} = e^{\boldsymbol{X}+\boldsymbol{Y}+\frac{1}{2}[\boldsymbol{X},\boldsymbol{Y}]+\frac{1}{12}[\boldsymbol{X},[\boldsymbol{X},\boldsymbol{Y}]]+\cdots} \tag{2.104}$$

Return operators for different time periods do not generally commute. If by order of magnitude $\langle\boldsymbol{R}_t\rangle \sim \sigma + \frac{\epsilon}{2}$, then commutator

$\langle[\boldsymbol{R}_1, \boldsymbol{R}_2]\rangle \sim \sigma^2 + \frac{\epsilon^2}{4}$. Accumulating over T periods the total error is $\sim T\left(\sigma^2 + \frac{\epsilon^2}{4}\right)$ and for typical values of σ and ϵ can be neglected compared to the other terms. With this precision we can say that the return operators add over time:

$$\boldsymbol{R} \approx \boldsymbol{R}_1 + \boldsymbol{R}_2 + \ \dots + \ \boldsymbol{R}_T \tag{2.105}$$

Therefore, if single-period return operator is described by Eq. (2.19), the return operator corresponding to T periods is:

$$\boldsymbol{R} \approx \mathbf{1} z\sqrt{T} + \frac{1}{2}\sqrt{\frac{T}{\tau}}\begin{pmatrix} \xi & \kappa \\ \kappa^* & -\xi \end{pmatrix} \tag{2.106}$$

Practically, in the context of trading we do not want to refer to the last price of the period preceding the initial and want to start counting from the already formed new price level. For this we must set $z_1 = 0$ and therefore:

$$\boldsymbol{R} \approx \mathbf{1} z\sqrt{T - \tau} + \frac{1}{2}\sqrt{\frac{T}{\tau}}\begin{pmatrix} \xi & \kappa \\ \kappa^* & -\xi \end{pmatrix} \tag{2.107}$$

As we remember, the high-low bars are the extension of spread into timeframes involving multiple trades. Their relative height h_r can now be computed as

$$h_r = 2z\sqrt{T - \tau} + \epsilon\sqrt{\frac{T}{\tau}} \tag{2.108}$$

Root-mean-square bar h is then equal[24]

---

[24] Same notation is used now for relative height as for absolute height in PART 1.

$$h = \sqrt{\langle h_T^2 \rangle} = \sqrt{4\sigma_z^2(T-\tau) + \epsilon^2 \frac{T}{\tau}} \tag{2.109}$$

for $T \geq \tau$. Introducing $T' = T - \tau$ to start counting from the first trade, Eq. (2.109) can be written as

$$h(T') = \sqrt{\epsilon^2 + (\epsilon^2 + 4\sigma_z^2\tau)\frac{T'}{\tau}} \tag{2.110}$$

or for semi-height

$$\frac{h(T')}{2} = \sqrt{\left(\frac{\epsilon}{2}\right)^2 + \left(\frac{\epsilon}{2}\right)^2 \frac{T'}{\tau} + \sigma_z^2 T'} \tag{2.111}$$

Taking the limit cases, we have

$$h(T' = 0) = \epsilon \tag{2.112}$$

$$h(T' \gg \tau) \approx 2\sqrt{\sigma_z^2 + \frac{\epsilon^2}{4\tau}}\sqrt{T'} = 2\sigma_\tau\sqrt{T'} \tag{2.113}$$

This goes along with Eqs. (1.8) and (1.9) and provides the answer to the question posed there and demonstrated in FIGURE 1.9: how does price uncertainty transition from constant spread to square-root behavior? Factor $\sigma_\tau = \sqrt{\sigma_z^2 + \frac{\epsilon^2}{4\tau}}$ plays the role of volatility related to time interval τ. In classical case, when $\epsilon = 0$ this factor reduces to σ_z. Finally, we can rewrite Eq. (2.110) in a convenient form as

$$\frac{h(T')}{2} = \sqrt{\left(\frac{\epsilon}{2}\right)^2 + \sigma_\tau^2 \frac{T'}{\tau}} \tag{2.114}$$

2.15 Price impact

In the previous section T was a free parameter. In the context of trade execution it becomes fixed. For an order of size Q that is being executed by submitting child orders with order size flow v (units/minute), time horizon $T = \frac{Q}{v}$. Question arises, what price impact can be expected from such operation?

Remember that for a single period of multistep evolution we have:

$$\frac{\delta s}{s} = z + \frac{\epsilon}{2} I \tag{2.115}$$

except the first step, in which $z = 0$. From here price impact of an elementary trade filling in a single period can be calculated by taking $I = 1$ or $I = -1$ depending on order direction, since execution is definite:

$$L_\tau = \pm\frac{\epsilon}{2} \tag{2.116}$$

Term z is not execution-dependent but must be preserved in the formula. For a lengthy execution spread must be replaced by the root-mean-square of the bar height corresponding to timeframe T:

$$L = \pm\sqrt{\langle\left(\frac{h}{2}\right)^2\rangle} \tag{2.117}$$

and again, the sign depends on order direction, which could be written as $\frac{Q}{|Q|}$. This factor will be implied but omitted in the following calculations in order not to overburden them.

Single-period return operator, corresponding to execution with flow v is:

$$\boldsymbol{R}_\tau = \boldsymbol{1}z + \frac{1}{2}\begin{pmatrix} \xi & \kappa - i\dfrac{2v}{\rho} \\ \kappa^* + i\dfrac{2v}{\rho} & -\xi \end{pmatrix} \tag{2.118}$$

To maintain validity of the coupled-wave model, order flow must satisfy $\frac{|v|\tau}{\rho} \sim \epsilon$. This makes the arguments of the previous section about the commutator hold right. Unlike other elements of the return operator, order stream is not random and therefore accumulates. The resulting return operator over time T is:

$$\boldsymbol{R}_T = \boldsymbol{1}z\sqrt{T-\tau} + \frac{1}{2}\begin{pmatrix} \xi\sqrt{\dfrac{T}{\tau}} & \kappa\sqrt{\dfrac{T}{\tau}} - i\dfrac{2vT}{\rho} \\ \kappa^*\sqrt{\dfrac{T}{\tau}} + i\dfrac{2vT}{\rho} & -\xi\sqrt{\dfrac{T}{\tau}} \end{pmatrix} \tag{2.119}$$

Note that this is the same operator that would come up if a bulk Q-unit order was just added to Eq. (2.106) without regard for how it was executed.

From here the root-mean-square high-low bar is equal

$$h = 2z\sqrt{T-\tau} + \sqrt{\xi^2\frac{T}{\tau} + \kappa'^2\frac{T}{\tau} + \left(\kappa''\sqrt{\frac{T}{\tau}} - \frac{2vT}{\rho}\right)^2} \tag{2.120}$$

Using Eq. (2.117), we get:

$$L = \pm\sqrt{\sigma_z^2(T-\tau) + \left(\frac{\epsilon}{2}\right)^2\frac{T}{\tau} + \left(\frac{Q}{\rho}\right)^2} \tag{2.121}$$

Remembering that $\sigma_\tau^2 = \sigma_z^2 + \frac{1}{\tau}\left(\frac{\epsilon}{2}\right)^2$ and using $T' = T - \tau$, we can rewrite it as

$$L = \pm \sqrt{\left(\frac{\epsilon}{2}\right)^2 + \sigma_\tau^2 \frac{T'}{\tau} + \left(\frac{Q}{\rho}\right)^2} \tag{2.122}$$

Since a trade we are interested in takes time $T \gg \tau$ we get:

$$L \approx \pm \sigma_\tau \sqrt{\frac{T}{\tau} + \left(\frac{Q}{\rho \sigma_\tau}\right)^2} \tag{2.123}$$

Substituting the execution horizon $T = \frac{Q}{v} = \frac{|Q|}{|v|}$, we obtain:

$$L \approx \pm \sigma_\tau \sqrt{\left|\frac{Q}{v}\right| + \left(\frac{Q}{\rho \sigma_\tau}\right)^2}$$

or, bringing it to daily values with $T = \frac{|Q|}{V} \frac{V}{|v|}$

$$L \approx \pm \sigma_{1d} \sqrt{\frac{|Q|}{V} + \left(\frac{Q}{\rho \sigma_{1d}}\right)^2} \tag{2.124}$$

For small Q we have

$$L(small\ Q) \approx \pm \sigma_{1d} \sqrt{\frac{|Q|}{V}} \tag{2.125}$$

which is the rule-of-thumb formula described in Section 1.7.

For large Q we have

$$L(large\ Q) \approx \frac{Q}{\rho} \tag{2.126}$$

which is a natural result complying with definition of ρ. The requirement of linear behavior for large orders was pointed out in Section 1.7. We also have Eq. (2.126) for highly liquid securities.

Impact scaling with order size is shown in FIGURE 2.25. Liquid securities sustain smaller impact than the illiquids. For liquid securities, the transition from square-root to linear scaling occurs sooner.

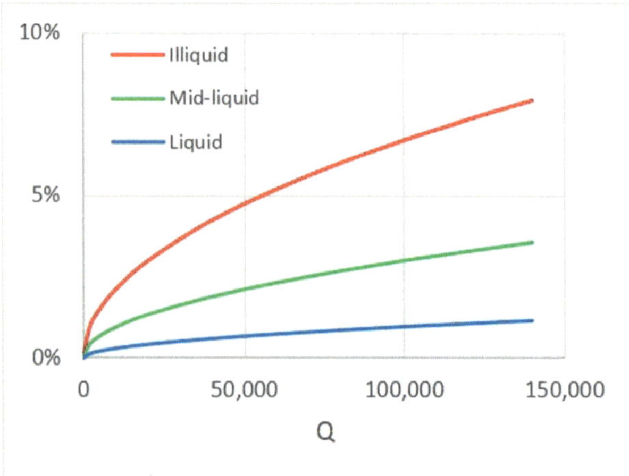

FIGURE 2.25 *Impact scaling with order size.*

Equation (2.124) suggests that price impact depends on the total order size and does not depend on how it is split into a stream. This is correct in this context since we disregarded the splitting profile by neglecting the commutators between return operators in Eq. (2.119). For larger order flow and non-uniform execution, the non-commutation of return operators becomes significant and must be taken into account.

It is implied that although statistical parameters of z, ξ, and κ, as well as ρ are taken constant here, they may vary in different situations. For example, market response can be included in the return operator as explained in Section 2.13. The end result is path dependent, which allows optimization in terms of finding the $v(t)$ that minimizes the cost of trade.

2.16 Risk management for illiquid securities

2.16.1 Market risk components

Market risk refers to possibility of financial losses due to movements in prices. Usually, market risk is quantified using the last price or closing price of the trading day. That works for liquid securities, which have a reasonably well-defined price at any time. Illiquid securities do not have a readily available price and require different approach. This is particularly true about fixed income securities, which commonly have large spread and low volatility. Presence of even one such instrument in the portfolio can impair the validity of risk model for the entire portfolio.

Coupled-wave model deals with illiquidity by decomposing market risk into three components (Sarkissian, 2016):

1. Diffusion risk – risk due to changes in overall price level. This risk is induced by the uncertainty of variable z in (2.47).

2. Spread uncertainty risk – arises due to uncertainty of the value of spread. Source of this risk is the uncertainty of ξ and κ in (2.48). Spread can change its value by multiples as can be seen in FIGURE 2.2.

3. Execution uncertainty – is the uncertainty of the next trade price, which can happen at either bid or ask level and which happens according to Eq. (2.51)

Each of these factors produces its own price distribution if other factors are eliminated. An example of this is shown in FIGURE 2.26. Only when all factors work together, do they produce a complete picture that includes effects associated with illiquidity.

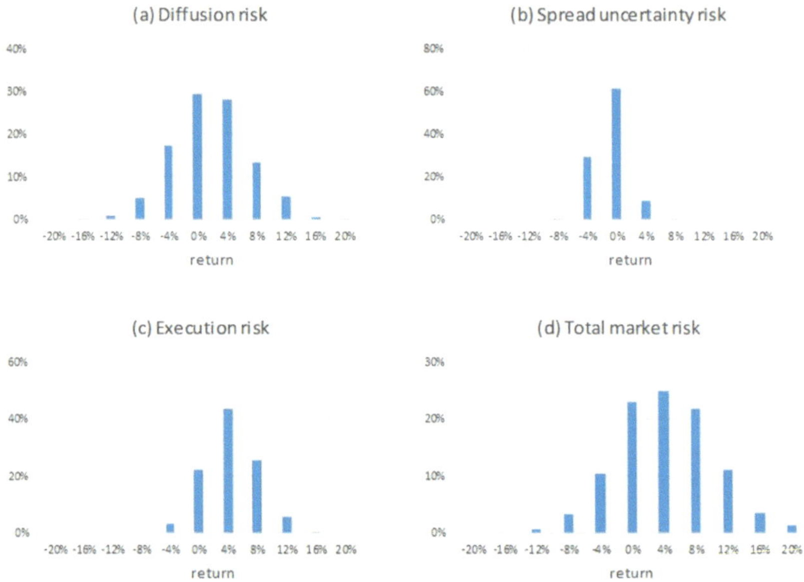

FIGURE 2.26 *Risk components and their combined effect.*

2.16.2 Liquidation losses and residual positions

A common measure of market risk is Value-at-Risk (VaR). Referring to a time horizon, it measures the size of losses that will not be exceeded with certain probability within the time horizon. VaR of liquid portfolios is based on mark-to-market pricing.

When portfolio contains illiquid securities, their risk assessment has to reflect the discount associated with spread, which can be done using the coupled-wave model. Same relates to large positions that affect market price in the course of their liquidation.

Consider a model long-only portfolio made of poorly correlated equities of various liquidity, shown in FIGURE 2.27. We want to know what maximum losses this portfolio may incur with a 95% confidence level within a 10-day time horizon.

Volatility and liquidity of individual positions are shown in FIGURE 2.28 in columns HV (historical volatility) and $Avg.V$ (Average daily volume) respectively. Ratio of position to daily volume is shown in the rightmost column Q/V. Many positions are liquid enough for a 10-day period. Some positions can be liquidated but not without a significant impact. There are also those, that will not liquidate within the time horizon.

According to mark-to-market calculations, the 95% 10-day VaR of this portfolio is $5,190,676, which is 3.07% of NAV. The assumption was that the entire portfolio was liquidated at the market prices.

In the meantime, if liquidity factor is taken into account, maximum realized loss for the liquidated part of portfolio would be $9,384,776 and there would remain a $43,673,721 residual portfolio of securities that could not be liquidated. Due to price impact and illiquidity, real losses will exceed mark-to-market VaR estimate by over 2 times. These are hidden risks, that if overlooked can lead to significantly underestimated risk picture.

Ticker	Quantity	Cur. price	Position
Security 1	393,202	31.08	12,220,718
Security 2	120,038	101.80	12,219,869
Security 3	73,978	165.12	12,215,247
Security 4	596,054	20.50	12,219,107
Security 5	47,657	256.36	12,217,348
Security 6	165,627	73.77	12,218,303
Security 7	633,466	19.29	12,219,560
Security 8	668,810	18.27	12,219,159
Security 9	39,855	306.18	12,202,804
Security 10	48,221	253.53	12,225,470
Security 11	427,230	11.00	4,699,530
Security 12	274,574	17.12	4,700,707
Security 13	80,088	58.68	4,699,564
Security 14	428,452	10.97	4,700,118
Security 15	390,664	12.03	4,699,688
Security 16	94,751	49.58	4,697,755
Security 17	517,657	9.08	4,700,326
Security 18	358,234	13.12	4,700,030
Security 19	132,258	35.54	4,700,449
Security 20	46,624	100.75	4,697,368
TOTAL			**169,173,119**

FIGURE 2.27 *Model equity portfolio with mixed liquidity.*

Ticker	Quantity	Cur. price	Position	HV	Ind. VaR	Avg. V	Q/V
Security 1	393,202	31.08	12,220,718	2.7%	1,746,849	335,082	1
Security 2	120,038	101.80	12,219,869	1.8%	1,138,706	434,926	0
Security 3	73,978	165.12	12,215,247	1.1%	710,719	4,032	18
Security 4	596,054	20.50	12,219,107	1.3%	818,836	2,471	241
Security 5	47,657	256.36	12,217,348	1.4%	893,741	497,311	0
Security 6	165,627	73.77	12,218,303	1.5%	943,353	1,354,390	0
Security 7	633,466	19.29	12,219,560	1.4%	907,433	32,848	19
Security 8	668,810	18.27	12,219,159	1.3%	857,337	43,424	15
Security 9	39,855	306.18	12,202,804	1.0%	633,718	148	269
Security 10	48,221	253.53	12,225,470	1.4%	885,184	194,892	0
Security 11	427,230	11.00	4,699,530	3.0%	747,839	726,023	1
Security 12	274,574	17.12	4,700,707	2.6%	633,922	34,013	8
Security 13	80,088	58.68	4,699,564	1.6%	387,295	1,516	53
Security 14	428,452	10.97	4,700,118	4.1%	1,013,473	26,626	16
Security 15	390,664	12.03	4,699,688	2.8%	675,534	1,586,834	0
Security 16	94,751	49.58	4,697,755	2.3%	574,291	6,397,226	0
Security 17	517,657	9.08	4,700,326	4.4%	1,087,634	121,698	4
Security 18	358,234	13.12	4,700,030	2.6%	645,409	45,556	8
Security 19	132,258	35.54	4,700,449	1.5%	364,402	1,925,982	0
Security 20	46,624	100.75	4,697,368	5.3%	1,309,094	755,594	0
TOTAL			**169,173,119**				
10d VaR-95%, $					**5,190,676**		
10d VaR-95%, %					**3.07%**		

FIGURE 2.28 *Liquidity and volatility overview. Ratio of position to daily volume is shown in the right column Q/V. Many positions are liquid enough for a 10-day period. Some positions can be liquidated but not without a significant impact. There are also those, that will not liquidate within the time horizon.*

Ticker	Quantity	Cur. price	Filling, #	Filling, %	Disc - exp	Disc - 95%	Rzd. loss	Max. rzd. loss (95%)
Security 1	393,202	31.08	393,202	100%	0.59%	6.05%	72,102	739,353
Security 2	120,038	101.80	120,038	100%	0.12%	4.94%	14,664	603,662
Security 3	73,978	165.12	40,323	55%	2.21%	4.97%	147,143	330,906
Security 4	596,054	20.50	0	0%	0.00%	0.00%	0	0
Security 5	47,657	256.36	47,657	100%	0.00%	3.60%	0	439,825
Security 6	165,627	73.77	165,627	100%	0.06%	4.00%	7,331	488,732
Security 7	633,466	19.29	328,484	52%	4.32%	7.83%	273,735	496,144
Security 8	668,810	18.27	434,242	65%	4.28%	7.80%	339,558	618,821
Security 9	39,855	306.18	0	0%	0.00%	0.00%	0	0
Security 10	48,221	253.53	48,221	100%	0.25%	4.98%	30,564	608,828
Security 11	427,230	11.00	427,230	100%	0.60%	7.84%	28,197	368,443
Security 12	274,574	17.12	274,574	100%	5.99%	13.23%	281,572	621,904
Security 13	80,088	58.68	20,000	25%	2.81%	6.10%	32,978	71,590
Security 14	428,452	10.97	428,452	100%	8.38%	16.22%	393,870	762,359
Security 15	390,664	12.03	390,664	100%	1.93%	8.75%	90,704	411,223
Security 16	94,751	49.58	94,751	100%	0.03%	6.59%	1,409	309,582
Security 17	517,657	9.08	517,657	100%	5.52%	17.45%	259,561	819,978
Security 18	358,234	13.12	358,234	100%	6.49%	14.65%	305,032	688,554
Security 19	132,258	35.54	132,258	100%	0.01%	4.04%	470	189,898
Security 20	46,624	100.75	46,624	100%	0.22%	17.35%	10,198	814,974
TOTAL RZD LOSS							**2,289,089**	
TOTAL MAX RZD LOSS								**9,384,776**

FIGURE 2.29 *Liquidation results.*

RESIDUAL PORTFOLIO			
Ticker	Res quantity	Cur. price	MTM
Security 3	33,655	165.12	5,557,183
Security 4	596,054	20.50	12,219,107
Security 7	304,982	19.29	5,883,106
Security 8	234,568	18.27	4,285,559
Security 9	39,855	306.18	12,202,804
Security 13	60,088	58.68	3,525,964
TOTAL			**43,673,721**

FIGURE 2.30 *Residual portfolio.*

PART III

QUANTUM THEORY OF PRICE FORMATION

3.1 Linking quantum and diffusion

Previous chapter dealt with price formation when market depth could be neglected, so that only two price levels – the best bid and the best ask – mattered. When coupled-wave approximation is inapplicable one has to solve the general propagation equation Eq. (2.53). That equation must involve as many price levels as may be important over the time horizon.

Unlike interactions in the particle world, interactions in finance are non-local. A large order hanging in the depth of order book will have an effect on price formation with the same immediacy as a small order at the top of the book. Any changes with the large order will have an immediate effect too. For example, its cancellation may result in price piercing through the level briefly or for good. Nonlocality of interactions is an essential difference between mechanics of prices and elementary particles.

With the dynamic coupling method described in the previous part price formation can often be approximated by a local process. After all, the majority of price formation does occur in the top levels of order book. When many price levels are taken into account, matrices representing operators can become so large that they can be approximated by differential operators[25]. The resulting differential equation governs evolution of probability amplitude in "coordinate representation". In this chapter we are going to refer to log-price as the coordinate: $x = ln(s/s_0)$.

Propagation equation does not have to be the same as Schrodinger equation. It has to be obtained from the matrix equation (2.53) as it applies to securities and markets without presuming any similarities with quantum mechanics operating with elementary particles. If this derivation is skipped, then important factors may be missed, and physical meaning of variables will be lost.

As a more general framework, we should expect the propagation equation in coordinate-representation to comply with the correspondence principle. We saw how this happens in the coupled-wave model. In continuous coordinate representation this means that for liquid securities the equation for probability distribution must turn into the diffusion equation:

$$\frac{\partial p}{\partial t} = K \frac{\partial^2 p}{\partial x^2} \tag{3.1}$$

with an initial condition

$$p(x, 0) = \delta(x)$$

Here K is the diffusion constant. This equation's solution is:

$$p(x, t) = \frac{1}{\sqrt{4\pi K t}} e^{-\frac{x^2}{4Kt}} \tag{3.2}$$

[25] Textbooks usually teach it the other way around – that finite difference operators approximate differential operators.

It is a Gaussian distribution that widens with time as its square root: $w \sim \sqrt{t}$, which is the familiar behavior of securities at large time scale $t \gg \tau$. On the other hand, we must expect that in the opposite extreme when $t \ll \tau$ the width of probability distribution must not scale at all: $w = const.$

Schrodinger equation too allows solutions in form of Gaussian distribution whose width does not scale for $t \ll \tau$. However, at large time scale it starts growing with time linearly and not as a square root: $w \sim t$. This raises numerous questions. How do these behaviors match? How does a quantum behavior transform into a square-root diffusion behavior? Is it possible at all to transform quantum propagation equation into the diffusion equation?

As the propagation equation will be derived later in this chapter, let us see how this transformation would work for the Schrodinger equation. Starting with a 1-dimensional time-dependent Schrodinger equation for a particle of mass m in a potential U

$$i\hbar \frac{\partial \psi}{\partial t} = -\frac{\hbar^2}{2m} \frac{\partial^2 \psi}{\partial x^2} + U\psi \tag{3.3}$$

and substituting

$$\psi = \eta e^{i\frac{\phi}{\hbar}} \tag{3.4}$$

we can get the following equation for probability density $p = \eta^2$

$$\frac{\partial \eta^2}{\partial t} + \frac{\partial}{\partial x}\left(\frac{1}{m}\frac{\partial \phi}{\partial x}\, \eta^2\right) = 0 \tag{3.5}$$

This equation has an illustrative meaning. In classical limit factor $v = \frac{1}{m}\frac{\partial \phi}{\partial x}$ is the particle's classical velocity. Therefore, Eq. (3.5) is a continuity equation for probability flow, which may look more familiar if written as

$$\frac{\partial p}{\partial t} + \frac{\partial (vp)}{\partial x} = 0 \tag{3.6}$$

This equation shows that probability density moves around in space like a particle with speed v, as shown in FIGURE 3.1. Unlike the diffusion equation, which operates with particle coordinate, quantum mechanics operates with localized probability density and is therefore unable to literally produce the diffusion equation for the coordinate. It can, however, produce the diffusion equation for the center of probability density.

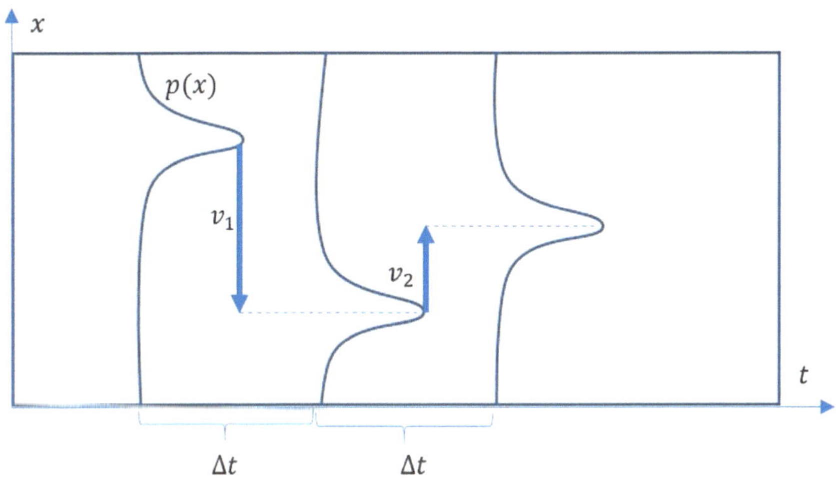

FIGURE 3.1 *Transition from quantum format to stochastic limit.*

Transition from quantum format to stochastic limit corresponds to applying Eq. (3.6) to a particle moving under a Brownian motion. Velocity of a Brownian particle is random from step to step but is constant within the time interval between collisions. For that matter, probability density $p(x, t + \Delta t) = p(x - v\Delta t, t)$. After each collision, center of probability distribution x_0 shifts randomly by an amount $v\Delta t$ while its shape is conserved. Indeed, writing $p(x, t)$ as the convolution of the particle's probability distribution relative to the center $\tilde{p}(x - x_0)$ and the center's probability distribution $y(x_0)$

$$p(x,t) = \int_{-\infty}^{\infty} \tilde{p}(x - x_0) y(x_0, t) dx_0 \tag{3.7}$$

we can show that the particle's localization center satisfies the diffusion equation:

$$\frac{\partial y}{\partial t} = \frac{v^2 \Delta t}{2} \frac{\partial^2 y}{\partial x_0^2} \tag{3.8}$$

As a result, after numerous collisions $p(x,t)$ widens in time even though the particle stays localized in space. With time the diffusion width takes over the particle size and becomes the only significant factor. This is how transition from quantum format to stochastic limit takes place.

3.2 Return spectrum

Multilevel return operator is represented by $N \times N$ Hermitian matrix. In the simplest setup its matrix elements are random, and their means and standard deviations are the parameters of the operator.

The return spectrum $\{r_n\}$ is determined from secular equation:

$$\det \|R_{mn} - r_n \delta_{mn}\| = 0 \tag{3.9}$$

If a security has a definite return value, then all diagonal elements of the return operator would equal to that return and all off-diagonal elements would equal to zero:

$$R_{mn} = r\,\delta_{mn}$$

Off-diagonal elements correspond to market depth. That depth is finite, so only a limited number of off-diagonal levels can be included. When interaction is limited to price levels adjacent to the current level the return operator is represented by a tridiagonal matrix:

$$R_{nn} \sim N(0, \sigma_\xi)$$

$$R_{n+1,n} \text{ and } R_{n-1,n} \sim CN(0, \sigma_\kappa)$$

This is the nearest neighbor approximation. Even though market depth is important, oftentimes its effect can be incorporated into the top levels as explained in the previous PART.

Fat tails of probability distribution are produced by distant off-diagonal elements:

$$R_{n+m,n} \text{ and } R_{n-m,n} \sim CN(0, \sigma_\kappa)$$

When N is large eigenvalues begin to densely populate the spectrum. Then probability distribution of returns can be written as

$$\chi(r) = \frac{1}{N}\sum_{n=1}^{N} \delta(r - r_n) \tag{3.10}$$

and can be approximated by a smooth curve. To find the return operator for a particular security, its parameters must be calibrated to make the spectrum match the observed return distribution $p(r)$. One can use the weighted least square criterion to assign weight to the tails:

$$\sum_{i=1}^{M} w_i |p(r_i) - \chi(r_i | \sigma_0, \sigma_1, \sigma_2 \ ...)|^2 = min \qquad (3.11)$$

Return spectrum calibrated to a probability distribution of daily returns for WMT ticker is shown in FIGURE 3.2.

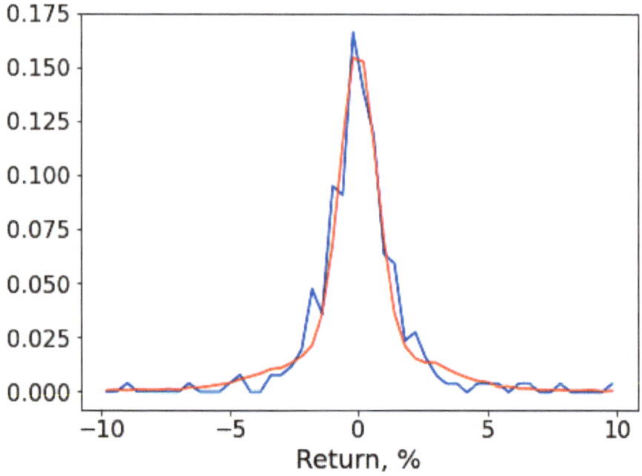

FIGURE 3.2 *Probability distribution of returns and calibrated return spectrum. Non-Gaussian features are captured by the long-range interactions between levels.*

3.3 Multilevel quantum dynamics

Evolution equation expressed through the return operator reads:

$$i\frac{\partial\psi}{\partial t} = \boldsymbol{R}\psi \qquad (3.12)$$

If $\boldsymbol{R}$ were a constant operator, ψ would have evolved smoothly and area of price localization would have gradually dispersed with time. FIGURE 3.3 shows this process when $\boldsymbol{R}$ equals the Schrodinger kinetic energy operator $\boldsymbol{R} = -\frac{\partial^2}{\partial x^2}$. Initial Gaussian wave packet would disperse while preserving its Gaussian shape, FIGURE 3.4 (left). In the process of dispersion, the peak probability would diminish, and probability of distant prices would gradually build up, FIGURE 3.4 (right). In the long run, width of the distribution would grow linearly FIGURE 3.5.

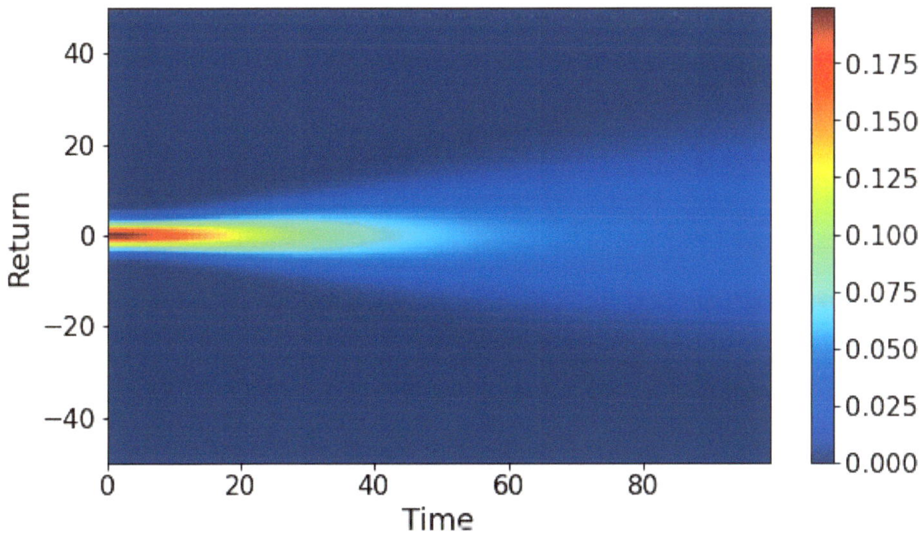

FIGURE 3.3 *Dynamics of probability distribution with constant return operator.*

With random return operator smooth evolution of probability amplitude is impossible. Instead, probability amplitude takes an irregular but localized shape that is made of numerous smaller probability formations. This dynamics is shown in FIGURE 3.6. Initial and final profiles are shown in FIGURE 3.7 (left).

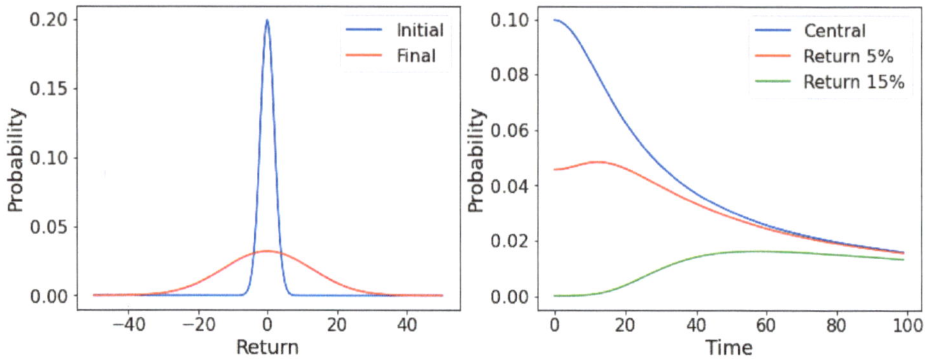

FIGURE 3.4 *Left: Initial and final probability distribution. Right: Probability dynamics with different price offsets.*

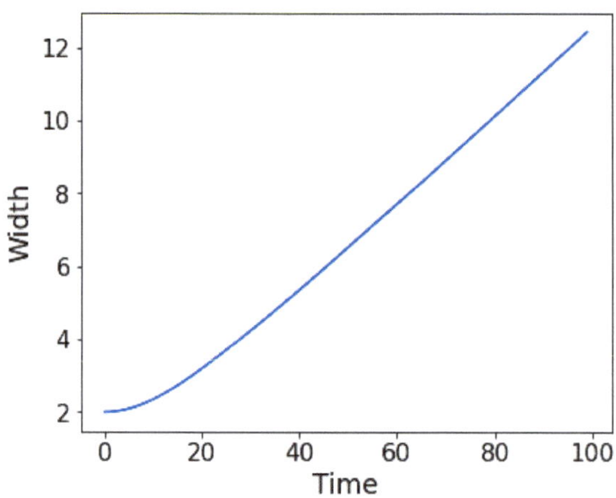

FIGURE 3.5 *Growth of the width of probability distribution under the constant differential return operator.*

Such character of probability distribution arises from irregular interference of different spectral components of the probability amplitude and has the same physical nature as the speckle pattern that occurs in diffuse reflections of monochromatic light, such as shown in FIGURE 3.8. Speckle patterns can be observed by reflecting laser light from paper, white paint, or any other rough surface. Similarity between erratic probability distribution and speckle pattern is genuine and not just by abstract analogy. Size of probability localization elements – the specklons – corresponds to the inhomogeneity of the matrix elements of the return operator.

Evolution of probability at the central and offset prices is shown in FIGURE 3.7 (right). Substantial level of probability may occasionally build up at any return level – big and small. This behavior can lead to adverse price moves known as the "Black Swans"[26].

Under random return operator dispersion slows down and turns from linear into square-root law. Its comparison to the benchmark from FIGURE 3.5 is shown in FIGURE 3.9.

Average over all possible evolution paths produces smooth dynamics, shown in FIGURE 3.10 and FIGURE 3.11. Ensemble averaging takes into account all possibilities and is important in pricing and risk evaluation problems. In reality evolution occurs along only one realization path with its erratic behavior.

---

[26] (Taleb, 2010)

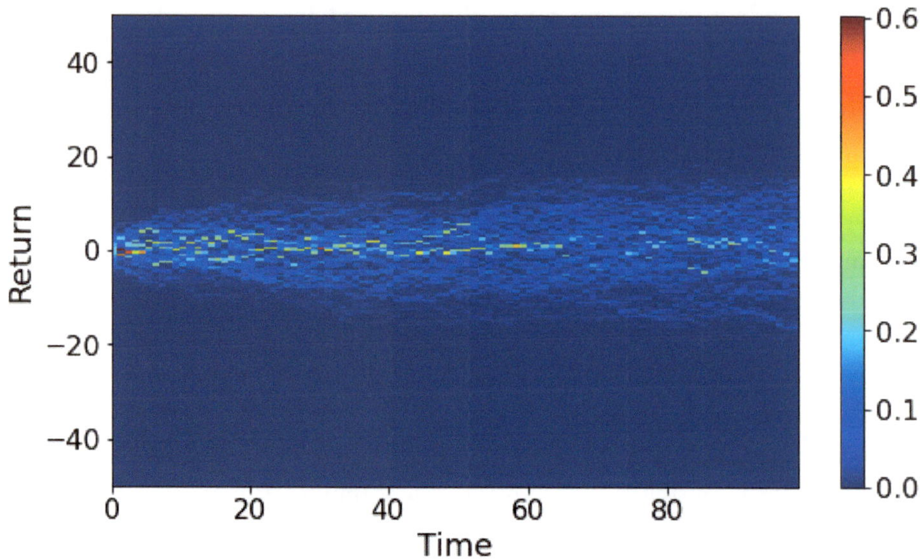

FIGURE 3.6 *Dynamics of probability distribution in a single realization Initial distribution disperses occasionally forming probability spikes that can be seen in the picture.*

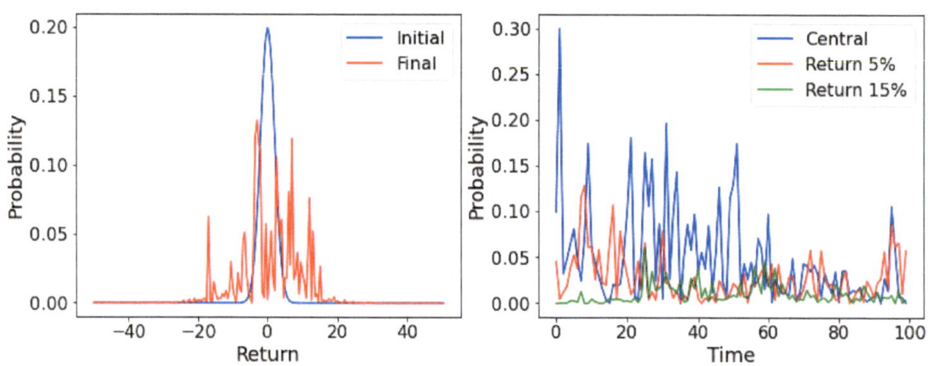

FIGURE 3.7 *Left: Initial and final probability distribution. Right: Probability dynamics with different price offsets.*

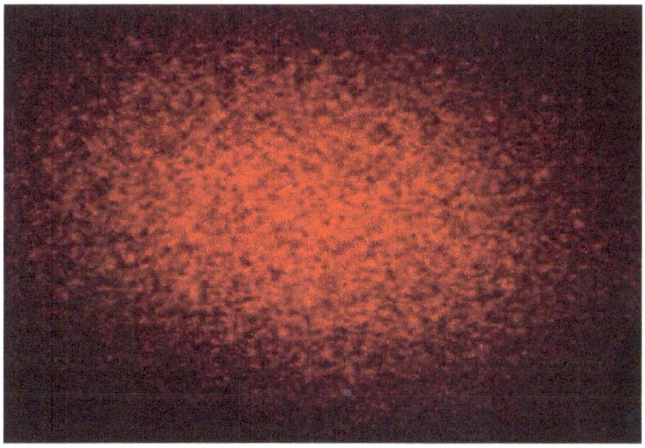

FIGURE 3.8 *A photograph of a speckle pattern. The pattern was formed by a laser beam scattered from plastic surface onto a wall. (This Photo by Unknown Author is licensed under CC BY-SA)*

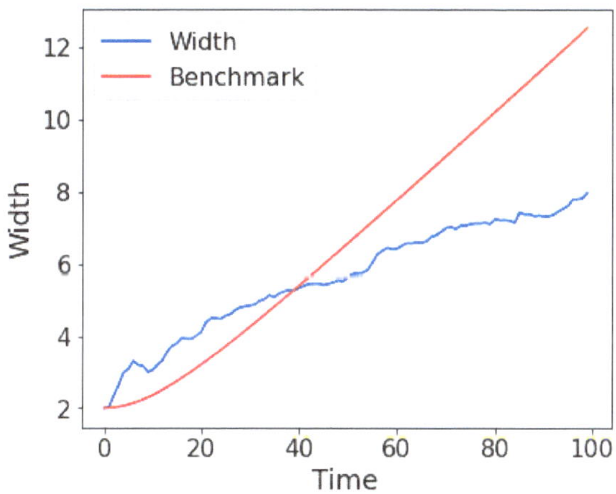

FIGURE 3.9 *Growth of the width of probability distribution under random return operator as compared to the benchmark from* FIGURE 3.5. *Probability localization slows down, and linear growth changes to square-root law.*

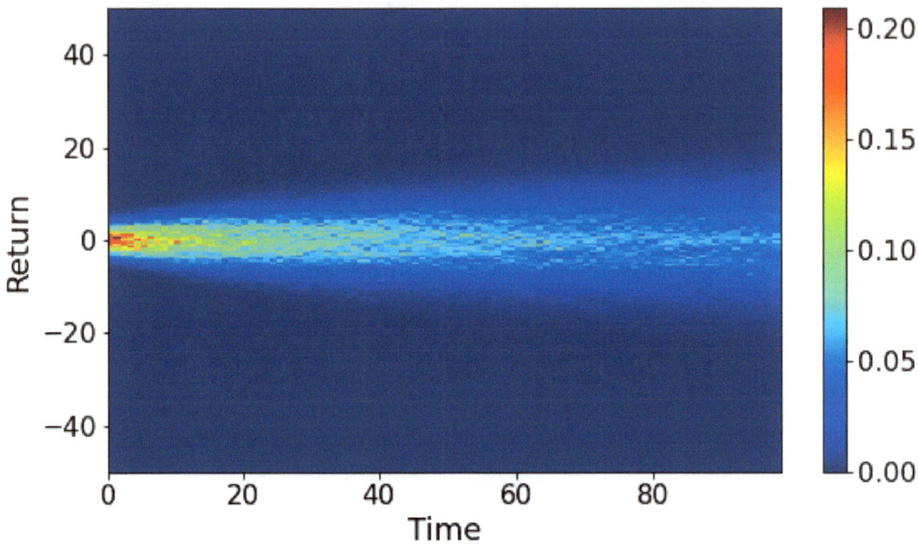

FIGURE 3.10 *Average dynamics of probability distribution.*

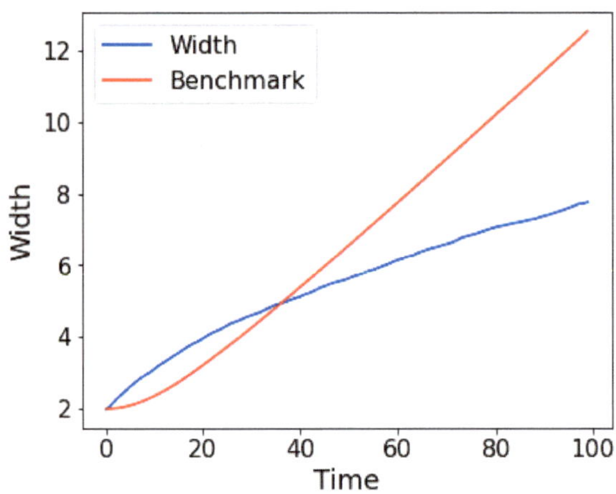

FIGURE 3.11 *Width of averaged probability distribution. Linear growth turns into a square-root law as diffusion takes over quantum effects in the long run.*

3.4 Coordinate representation – Market Maker's Equation

Now is the time to derive the state evolution equation in continuous coordinate representation. In quantum mechanics for elementary particles this is done using the correspondence principle, in which variables are replaced by their respective operators.

Finance lacks the ability for such transition. The only part that can be reliably used for correspondence principle is the diffusion equation, which is (a) written for probability distribution, and (b) does not contain spread. We must therefore rely of a different principle in derivation.

Such principle comes from considering microstructural processes leading to price dynamics and formation. The key to that is the coupling of price levels which is ultimately responsible for dynamic transfer of probability amplitude between levels (Sarkissian, 2016). Representing the dynamic equation Eq. (3.12) in the form that factors out the transition coefficients, we will be able to take the continuous limit to the differential form.

Let us write Eq. (3.12) in matrix form:

$$i\frac{\partial \psi_m}{\partial t} = \sum_n R_{mn}\psi_n \tag{3.13}$$

In the nearest neighbor approximation, return operator is a tridiagonal matrix:

$$R_{mn} = R_{m,m-1}\delta_{m-1,n} + R_{m,m}\delta_{mn} + R_{m,m+1}\delta_{m+1,n} \tag{3.14}$$

and therefore,

$$i\frac{\partial\psi_m}{\partial t} = R_{m,m-1}\psi_{m-1} + R_{m,m}\psi_m + R_{m,m+1}\psi_{m+1} \tag{3.15}$$

Since return operator is Hermitian, the following relation must hold:

$$R_{m,m+1} = R^*_{m+1,m}$$

First and last terms of Eq. (3.15) correspond to inflow of probability amplitude into level m from adjacent levels. The middle term contains outflow of probability amplitude from level m into adjacent levels as shown in FIGURE 3.12. Let us rewrite the propagation equation in a way that reflects it:

$$i\frac{\partial\psi_m}{\partial t} = r_{m,m-1}\psi_{m-1} - \left(r_{m-1,m} + r_{m,m} + r_{m+1,m}\right)\psi_m + \\ + r_{m,m+1}\psi_{m+1} \tag{3.16}$$

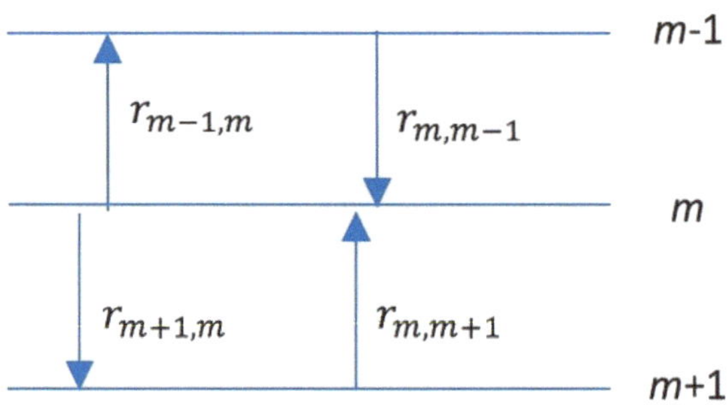

FIGURE 3.12 *Transitions between price levels.*

Here r_{nm} are transition coefficients from level m to level n. In the nearest-neighbor approximation they depend only on their number on the diagonal, m. Further, we will omit the coefficient $r_{m,m}$, which can be eliminated by a trivial phase transformation. We will bring it back in the final result. Let us set

$$r_{m,m-1} = b_{m-1} - \frac{a_{m-1}}{2}$$

$$r_{m,m+1} = b_{m+1} + \frac{a_{m+1}}{2}$$

From here coefficients a_m and b_m can be written as:

$$a_m = r_{m-1,m} - r_{m+1,m}$$

$$b_m = \frac{r_{m-1,m} + r_{m+1,m}}{2}$$

We can rewrite Eq. (3.16) in terms of coefficients a_m and b_m:

$$i\frac{\partial \psi_m}{\partial t} = \frac{1}{2}(a_{m+1}\psi_{m+1} - a_{m-1}\psi_{m-1}) +$$

$$+(b_{m+1}\psi_{m+1} - 2b_m\psi_m + b_{m-1}\psi_{m-1})$$

Recognizing the difference representations of the first and second derivatives in the right-hand side, let us set

$$a = \frac{1}{\tau}\frac{A}{\Delta x} \quad and \quad b = \frac{1}{\tau}\frac{B}{\Delta x^2}$$

and therefore:

$$i\tau\frac{\partial \psi}{\partial t} = V\psi + \frac{\partial(A\psi)}{\partial x} + \frac{\partial^2(B\psi)}{\partial x^2} \tag{3.17}$$

where the term $V\psi$ accounts for the previously omitted coefficient $r_{m,m}$. This is the general form of the evolution equation in coordinate representation. It is called the **Market Maker's Equation**.

Expanding the derivatives, we can write Eq. (3.17) more conveniently as

$$i\tau\frac{\partial\psi}{\partial t} = U\psi + D\frac{\partial\psi}{\partial x} + B\frac{\partial^2\psi}{\partial x^2} \tag{3.18}$$

where

$$U = V + \frac{\partial A}{\partial x} + \frac{\partial^2 B}{\partial x^2} \quad \text{and} \quad D = A + 2\frac{\partial B}{\partial x} \tag{3.19}$$

Market Maker's Equation consists of three right-hand terms, which can be labeled according to their basic function:

$U\psi$ – potential

$D\frac{\partial\psi}{\partial x}$ – drift

$B\frac{\partial^2\psi}{\partial x^2}$ – dispersion and focusing

As we see, the evolution equation is NOT the same as the Schrodinger equation. It has an additional term $D\frac{\partial\psi}{\partial x}$, which is generally not equal to zero at any given moment but can be equal to zero on average. It is missing from the Schrodinger equation (in the absence of external field) due to the fact that for elementary particles $D = 0$ precisely. Further, not only the potential $U(x,t)$, but also $D(x,t)$ and $B(x,t)$ can be functions of the coordinate and time. If Schrodinger equation were simply borrowed from quantum mechanics related to elementary particles, we would have missed these important details.

Values of A at points x_m are determined by the differential between coupling $r_{m-1,m}$ and $r_{m+1,m}$, and values of B reflect the degree of that coupling. Functions $A(x,t)$ and $B(x,t)$ are determined by market environment. They can also reflect trader's own trading strategy with its timing and reaction to price changes. Remembering that A and B run only one of many possible realizations, ensemble averaging is still in force to obtain the final probability distribution:

$$P(x,t) = \langle \left| \psi^{(r)} \right|^2 \rangle_r \tag{3.20}$$

where index r corresponds to realization.

3.5 Trading and price shift

The right-hand side of the Market Maker's Equation is nothing but differential form of the return operator. Just like in coupled-wave model, it would be good to see how it is related to the trade operator. Since trading is associated with price shifts, operator corresponding to a trade must be related to price shift operator.

Consider an action of a translational operator $\boldsymbol{U}_{\boldsymbol{\delta x}}$ that shifts the coordinate of probability amplitude by a small amount δx:

$$\boldsymbol{U}_{\boldsymbol{\delta x}} \psi(x) = \psi(x + \delta x) \tag{3.21}$$

Using expansion

$$\psi(x + \delta x) \approx \psi(x) + \frac{\partial \psi}{\partial x} \delta x \tag{3.22}$$

$\boldsymbol{U}_{\boldsymbol{\delta x}}$ can be written as

$$\boldsymbol{U}_{\boldsymbol{\delta x}} = 1 + \delta x \frac{\partial}{\partial x} \tag{3.23}$$

In a market with limit orders distributed uniformly on each side with order density ρ a trade of size q results in price shift $\delta x = \frac{q}{\rho}$. Therefore, operator

$$\frac{q}{\rho} \frac{\partial}{\partial x}$$

is associated with trading q units of security. Considering the fact that trade quantity operator must be Hermitian, and that the elementary price shift is $\delta x = \frac{1}{\rho}$, **trade operator** is:

$$\boldsymbol{T} = \frac{1}{i\rho}\frac{\partial}{\partial x} \tag{3.24}$$

Trade operator corresponds to trading 1 unit of security and has dimensions of $\left[\frac{1}{\#}\right]$. Its eigenfunctions are plane waves:

$$\psi = e^{i\rho\frac{x}{q}} \tag{3.25}$$

so that

$$\boldsymbol{T}\psi = \frac{1}{q}\psi \tag{3.26}$$

Trade operator relates to a single trade that fills immediately. Often instead of a single trade we have to deal with trade flow. **Inverse flow operator** is defined as

$$\boldsymbol{\Theta} = \tau\boldsymbol{T} = \frac{1}{i\theta}\frac{\partial}{\partial x} \tag{3.27}$$

Here $\theta = \frac{\rho}{\tau}$ is the **order density flow**. Inverse flow operator's eigenfunctions are the same as the trade operator's, but eigenvalues are $\frac{\tau}{q}$, which is the inverse of trade flow $\frac{q}{\tau}$:

$$\boldsymbol{\Theta}\psi = \frac{\tau}{q}\psi \tag{3.28}$$

Trade operator can serve as a base for constructing other operators. For example, operator $\frac{F}{i}\frac{\partial}{\partial x}$ with F representing the market cash volume corresponds to trade flow. It assumes the existence of actively traded markets generating cash volume F, whereas the trade operator relies only on the trade under consideration.

3.6 Financial uncertainty principle

Coordinate operator corresponds to the position of log-price and is represented simply by its value: $\boldsymbol{X} = x$. Commutator of price and inverse flow operators is:

$$[\boldsymbol{\Theta}, \boldsymbol{X}] = \frac{1}{i\theta} \tag{3.29}$$

Therefore, standard deviations of the two quantities must satisfy the uncertainty relation:

$$\sigma_X \, \sigma_\Theta \geq \frac{1}{2\theta} \tag{3.30}$$

This relation can be obtained by computing the function:

$$f(a) = \int_{-\infty}^{\infty} \left| ax\psi + \frac{d\psi}{dx} \right|^2 dx$$

and requiring $f(a) \geq 0$ for $\forall a$. Detailed proof of this can be found in most textbooks on quantum mechanics.

In example from Section 1.3 we used buy and sell orders for measurement. We saw that standard deviation of price was $\frac{\epsilon}{2}$, standard deviation of order size was m, and the process took time τ. In that market local order density was $\frac{M_{bid}+M_{ask}}{2\epsilon}$, since we had two buy and sell orders that were separated by the spread. Therefore, order size density is $\rho = \frac{M}{\epsilon}$ with average order size $M = \frac{M_{bid}+M_{ask}}{2}$. To make sure that trading does not penetrate the top levels we must have $m \leq M$. Therefore,

$$\sigma_X \sigma_\Theta = \frac{\epsilon}{2}\frac{\tau}{m} = \frac{1}{2\theta}\frac{M}{m} \geq \frac{1}{2\theta} \tag{3.31}$$

with $\theta = \frac{m}{\epsilon\tau}$.

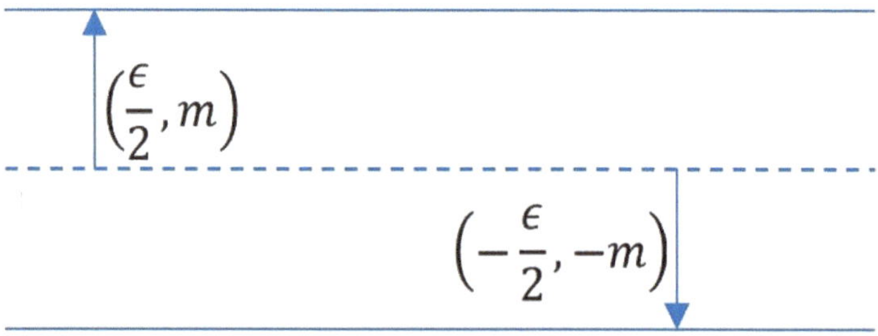

FIGURE 3.13 *Measurement results with buy and sell orders.*

The uncertainty relation shows that price accuracy can be improved by increasing the order density flow. That in turn can be achieved by supplying bigger and denser order size to the market and reducing the time between trades. All of this aggregates into the notion of market liquidity.

3.7 Quantum-diffusion transition

Unlike Schrodinger equation, each term in the right-hand side of the Market Maker's Equation does not have to be Hermitian on its own. The total of the terms has to be Hermitian, but each separate term is free of that requirement. Let us consider the case when both terms are Hermitian. In this case the Market Maker's Equation has to take the following form:

$$i\frac{\partial\psi}{\partial t} = iz(t)\frac{\partial\psi}{\partial x} + c(t)\frac{\partial^2\psi}{\partial x^2}, \tag{3.32}$$

where z and c are real. For simplicity we will set $\tau = 1$ for intermediate calculations and resurrect it in the final results.

Parameters $z(t)$ and $b(t)$ are piecewise constant in time and refresh after each time step:

$$z(t) = z_k, \text{ for } t_k \leq t < t_{k+1}$$

$$c(t) = c_k, \text{ for } t_k \leq t < t_{k+1}$$

with $z{\sim}N(0,\sigma_z)$ and $c{\sim}N(0,\sigma_c)$. Note, that unlike in Schrodinger equation, here c can be both negative and positive.

Under these conditions, let us consider evolution of a localized initial state with Gaussian distribution that has half-width w_0, so that the normalized probability amplitude is:

$$\psi(x,0) = \frac{1}{\sqrt[4]{2\pi w_0^2}} e^{-\frac{x^2}{4w_0^2}} \tag{3.33}$$

Obviously, $w_0{\sim}\epsilon$ by the order of magnitude.

Solving Eq. (3.32) for one realization over the interval Δt, we get:

$$\psi^{(i)}(x,\Delta t) = \frac{1}{\sqrt[4]{2\pi w^2}} e^{-\frac{\left(x+z^{(i)}\Delta t\right)^2}{4\left(w_0^2 - ic^{(i)}\Delta t\right)}} \tag{3.34}$$

Propagation over multiple steps yields:

$$\psi^{(i)}(x, N\Delta t) = \frac{1}{\sqrt[4]{2\pi w^2}} e^{-\frac{\left(x+\Delta t \sum_{k=1}^{N} z_k^{(i)}\right)^2}{4\left(w_0^2 - i\Delta t \sum_{k=1}^{N} c_k^{(i)}\right)}} \tag{3.35}$$

Taking it one step further, we have:

$$P^{(i)}(x, N\Delta t) = \frac{1}{\sqrt{2\pi w^2}} e^{-\frac{\left(x+\Delta t \sum_{k=1}^{N} z_k^{(i)}\right)^2}{2\left[w_0^2 + \frac{1}{w_0^2}\left(\Delta t \sum_{k=1}^{N} c_k^{(i)}\right)^2\right]}} \tag{3.36}$$

or

$$P^{(i)}(x, t) = \frac{1}{\sqrt{2\pi w^2}} e^{-\frac{\left(x+Z^{(i)}\right)^2}{2\left[w_0^2 + \left(\frac{C^{(i)}}{w_0}\right)^2\right]}} \tag{3.37}$$

where $Z^{(i)} = \Delta t \sum_{k=1}^{N} z_k^{(i)}$ and $C^{(i)} = \Delta t \sum_{k=1}^{N} c_k^{(i)}$. This distribution has half-width $w = \sqrt{w_0^2 + \frac{\sigma_c^2}{w_0^2} t}$.

Finally, taking the ensemble average over all possible realizations, we get:

$$P(x, t) = E_i\left[\left|\psi^{(i)}(x, t)\right|^2\right] = \frac{1}{\sqrt{2\pi w(t)^2}} e^{-\frac{x^2}{2w(t)^2}} \tag{3.38}$$

where the half-width is given by:

$$w(t) = w_0 \sqrt{1 + \left[\frac{\sigma_z^2}{w_0^2} + \left(\frac{\sigma_c}{w_0^2}\right)^2\right] t} \tag{3.39}$$

or

$$w(t) = \sqrt{w_0^2 + \frac{\sigma_c^2}{w_0^2} t + \sigma_z^2 t} \tag{3.40}$$

Comparing this to Eq. (2.111), we see that the following correspondence is in place: $w_0 = \frac{\epsilon}{2}$ and $\frac{\sigma_c^2 \tau}{w_0^2} = \left(\frac{\epsilon}{2}\right)^2$. Ultimately, we have the same formula:

$$w(t) = \sqrt{\left(\frac{\epsilon}{2}\right)^2 + \left(\frac{\epsilon}{2}\right)^2 \frac{t}{\tau} + \sigma_z^2 t} \tag{3.41}$$

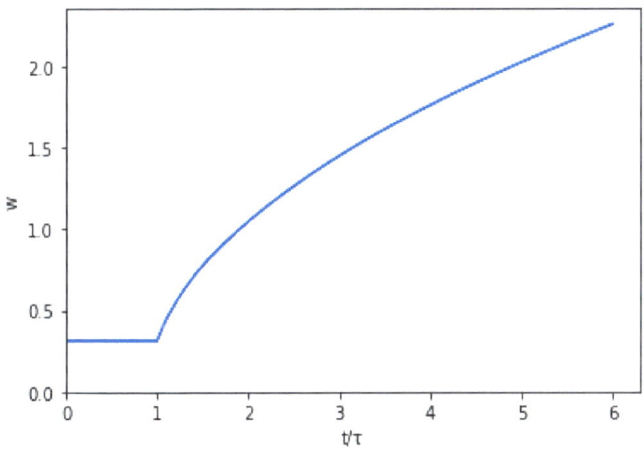

FIGURE 3.14 *Price uncertainty scaling with time.*

From here, parameter σ_c can be estimated through relation $\sigma_c\sqrt{\tau} = \left(\frac{\epsilon}{2}\right)^2$. Coefficient σ_z has the same meaning as in the coupled-wave model: it is the shift of the general price level. Given the average order flow imbalance $\delta v = \frac{\delta q}{\tau}$, it can be estimated as $\sigma_z\sqrt{\tau} \approx \frac{\delta q}{\rho}$.

Equation Eq. (3.40) unifies the quantum behavior at $t < \tau$ and diffusive square root behavior at $t \gg \tau$. At small time scale width is constant equal ϵ. At large time scale

$$w(t) \approx \sigma\sqrt{t}$$

which is the familiar diffusive behavior with $\sigma = \sqrt{\frac{\sigma_c^2}{w_0^2} + \sigma_z^2}$. Equation (3.40) fills the gap in FIGURE 1.9. The completed plot is shown in FIGURE 3.14.

3.8 Price stabilization and fair value

If a security has **fair value** x_0, its price should be localized around it. However, Market Maker's Equation does not contain an explicit reference to fair value. It is determined by the market and is inexplicitly embedded in functions $A(x,t)$ and $B(x,t)$. These functions not only determine the fair value, but also encode market reaction to deviations from it. No matter what input is given as the initial condition, that input will have to evolve into a stable distribution localized around the fair value. Let us find that distribution and the corresponding market conditions.

Since we are looking at the market that is in agreement regarding the fair value, $\frac{\partial}{\partial t} \equiv 0$. Functions $A(x)$ and $B(x)$ are subject to relevant restrictions that characterize securities at fair value. In a stable state the transition coefficients are symmetric: $r_{m-1,m} = r_{m,m-1}$, and therefore $A(x)$ and $B(x)$ must be real functions. Further, at $x = x_0$ transition coefficients in the opposite directions must be equal since there is no preference in price direction, so $r_{m-1,m} = r_{m+1,m}$. Therefore $A(x_0) = 0$.

Function $A(x)$ is the local imbalance of transition coefficients. To guarantee stability, $A(x)$ has to be a monotonously increasing or decreasing function of x: $\frac{\partial A}{\partial x} < 0$, or $\frac{\partial A}{\partial x} > 0$ depending on the sign of r.

Solution of Eq. (3.17) that satisfies these conditions is:

$$\psi(x) = \frac{1}{\sqrt{2\Delta x}\, cosh\left(\frac{x - x_0}{\Delta x}\right)} \tag{3.42}$$

with

$$A(x) = A_0\, tanh\left(\frac{x - x_0}{\Delta x}\right) \tag{3.43}$$

and $\Delta x = \frac{B}{A_0}$. The value of x_0 is determined by the market, depending on where $A(x_0) = 0$.

The corresponding probability distribution is:

$$p(x) = \frac{1}{2\Delta x} \frac{1}{cosh^2\left(\frac{x - x_0}{\Delta x}\right)} \tag{3.44}$$

For small price interval:

$$p(s) \approx \frac{1}{2\Delta s} \frac{1}{cosh^2\left(\frac{s - s_0}{\Delta s}\right)} \tag{3.45}$$

Taking imbalance as the difference between cumulative execution probabilities above and below the trading price, we have

$$I(x) = \int_x^{\infty} p(x')dx' - \int_{-\infty}^{x} p(x')dx' = -tanh\left(\frac{x - x_0}{\Delta x}\right) \tag{3.46}$$

The same result for imbalance can be obtained by considering ensembles of securities as will be shown in PART 4. Even though the fair value is fixed at x_0, it is only as definite as the spread Δx

allows it to be. Deviations of market price from x_0 will experience an imbalance with an appropriate sign, but it will not be strictly ± 1. This flexibility allows for price uncertainty.

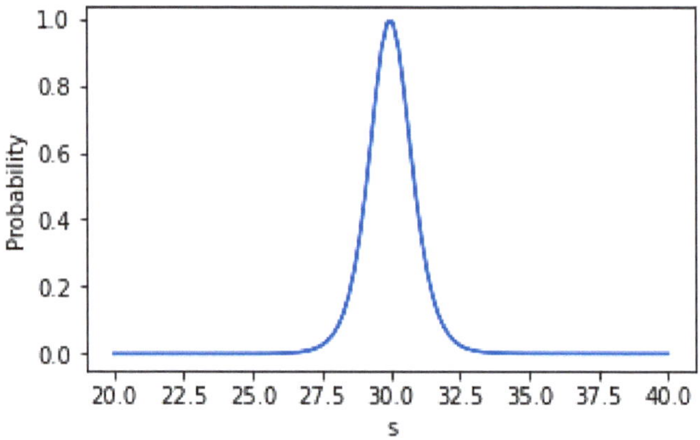

FIGURE 3.15 *Probability distribution of price in equilibrium market.*

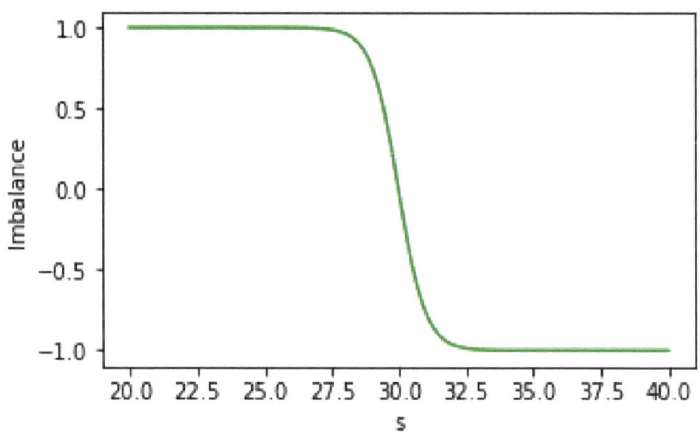

FIGURE 3.16 *Imbalance in equilibrium market.*

Assumptions that went into solution Eq. (3.42) correspond to a frozen market without trades. In this sense it is a stable solution as opposed to stationary solution, in which probability transfer can also occur due to trades. Such market can be described similarly, but with complex transition coefficients.

What makes the solution Eq. (3.42) possible? In stochastic models probability evolution can take only the form of dissipation, requiring the width of probability distribution to increase with time. In quantum framework probability evolution can combine dissipation and dispersion. It this case they work with equal strength but in opposite directions thus cancelling each other.

Equation (3.43) can be reversed in its interpretation. Since $A(x) \sim I \sim \delta q$, if a market is arranged in which $\delta q \sim -tanh\left(\frac{x-x_0}{\Delta x}\right)$, i.e. securities are bought or sold in quantities proportionate to the $tanh(x)$ profile, then price distribution will be described by Eq. (3.45). Market Maker's Equation (3.17) allows to find what conditions are necessary to obtain the desired probability distribution or to ensure its preferred dynamics.

3.9 Option pricing for illiquid securities

Trading options on underlying with limited liquidity poses multifaceted problems. Price gaps, non-Gaussian statistics and liquidation costs complicate the already complex picture of option valuation.

When the underlying has limited liquidity, option holder does not get the full intrinsic value at exercise or expiration. For example, buyer of an in-the-money vanilla call option will not receive the full $S-K$ difference. They will only receive $S-K-\delta$ adjusted for haircut δ that reflects price impact and depends on the notional. In turn, costs to call seller will raise by δ. Therefore, option valuation with illiquid underlying must be based on adjusted strike ranging from $K - \delta$ to $K + \delta$ depending on the direction and position netting.

This effect does not depend on which option pricing model is used. Given the notional amount liquidation costs can be calculated using the framework described in previous sections, for example the coupled-wave model. Buy and sell prices taking into account liquidity effects for a call are shown in FIGURE 3.17. The two prices reflect buyer's and seller's liquidity concerns and form the basis of option's bid-ask spread.

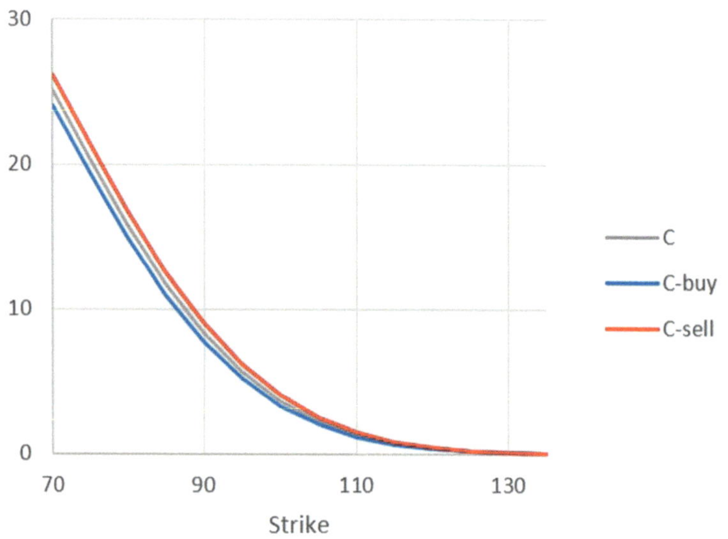

FIGURE 3.17 *Bid and ask prices for a call on illiquid underlying for different notional amounts.*

One of the difficult problems with options on illiquid underlying is computation of the Greeks. Since there is no readily available market and spread is large, the Greeks seem to have no definite value. Unless options are to be held to expiration, lack of clarity in estimation of the Greeks can pose a serious problem, making it difficult to hedge properly, inflating transaction costs due to frequent re-hedging and distorting risk profile. This is particularly important in situations when otherwise liquid underlying suddenly loses liquidity.

As long as some underlying market exists, even in form of indicative quotes, quantum models can be calibrated without assuming that trades are necessarily happening. Then, based on the calibrated underlying process, option prices can be computed according to their definitions.

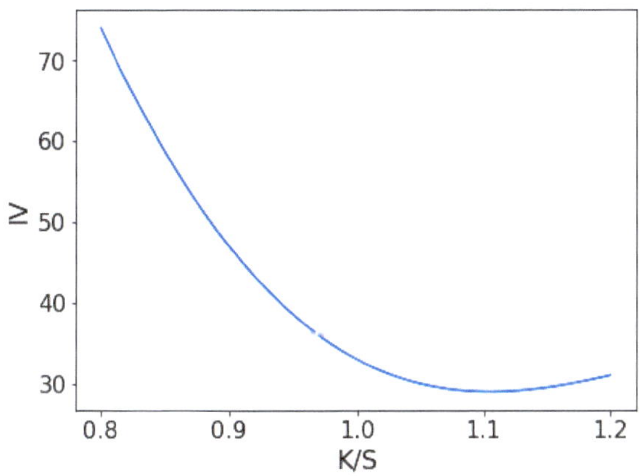

FIGURE 3.18 *Volatility skew of a call option priced under quantum framework. In this framework typical features of equity volatility surface appear naturally due to long-range off-diagonal elements in return operator.*

Since in quantum framework price process is generally non-Gaussian typical features of volatility appear in it naturally (Sarkissian, 2016). Volatility skew of a call option priced under quantum framework is shown in FIGURE 3.18 and volatility surface is shown in FIGURE 3.19. These features appear due to inclusion of spread into picture and due to long-range off-diagonal elements of return operator.

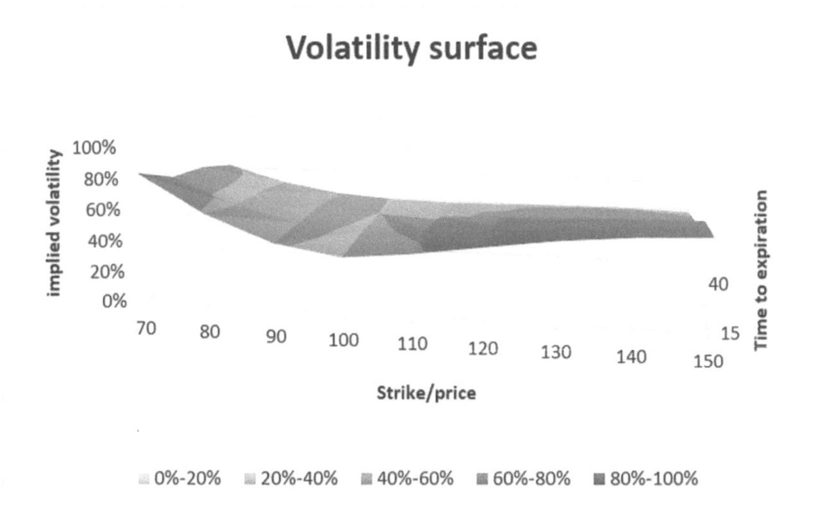

FIGURE 3.19 *Volatility surface of a call option priced under quantum framework. In this framework typical features of equity volatility surface appear naturally due to long-range off-diagonal elements in return operator.*

PART IV

STATISTICAL MECHANICS OF FINANCIAL MARKETS

4.1 Markets as statistical systems

Randomness of financial markets blurs the underlying laws connecting them to scientific principles. Scale of fluctuations in finance is so large that the laws can become simply unnoticeable. Common physical macro systems are made of 10^{18} particles or more. With so many particles their macro characteristics fluctuate by about 10^{-9} of their mean values ($\sim\frac{1}{\sqrt{N}}$).

Financial markets are made of much smaller number of elements. Number of securities in a particular market is measured in hundreds or thousands. For example, the 3 main US stock exchanges list just about 6,500 equities. Only 500 members of S&P 500 comprise over 70% of total market capitalization. Bond market counts about 30,000 issues from only a few thousand different issuers[27]. There are only 180 currencies in the world, 50 of them are pegged, and most independent ones have thin localized markets.

These numbers correspond to fluctuations of about a few percent. It takes only a few steps for the few percent to grow into the double-

[27] cbonds.com

digit range. With such numbers fundamental laws become very hard to notice. Yet, they may transpire in collective behavior if their effect is amplified. This can happen if a physical effect scales with the number of securities faster than fluctuations. Therefore, it makes sense to look at the markets as collections of securities and describe statistical properties of these collections.

Fluctuation scale is not the only difference of the markets from physical macro systems. Common physical systems are closed systems, which means that they can exist and evolve on their own without significantly affecting the outside world. To the contrary, financial markets are highly interconnected and have multiple layers. For instance, the same security may be listed on multiple exchanges, listing may be in different currencies or as depositary receipts, shares may be traded in electronic markets or over the counter, etc. Such granularity of market structure complicates it as a physical system.

Another difference relates to market dynamics. In our intuitive understanding markets are never at equilibrium. New information flows constantly, preventing the markets from reaching equilibrium. One may suggest that kinetics, rather than statistical mechanics could be a more appropriate framework to study the markets. While kinetics has its place, we will see that market equilibrium is reachable if it is reformulated.

Due to interaction with reservoir, parameters of common physical systems can be regulated externally. For instance, one can increase the temperature of a liquid crystal cell by applying heat. This cannot be done with the markets. Markets do not have an external world – they are the financial world. They do, however, have central banks and regulators that can influence market behavior to a certain extent.

These details complicate the study of financial markets as statistical systems. Yet, markets are statistical systems, and it is the purpose of this chapter to provide their description as such.

4.2 Ensemble view of the markets

The cornerstone of statistical mechanics is a **statistical ensemble**[28]. Statistical ensemble is a collection of all possible states of a system. It is a very useful notion due to its ability to facilitate convenient computation of statistical measures. The idea is that if you have the ensemble, then you do not have to wait until the system goes through its all states to compute its statistics. You can instead compute statistics over the ensemble since it already contains all possible states. That being said, the ability to do so is contingent on the system's ergodic properties, which will be addressed in the following sections.

Financial markets form ensembles of investment dollars distributed across securities (Sarkissian, 2016). When a security makes a 2% return, we can say that the investment dollars in that security were in a state with return 2%. If at the same time, another security drops 1%, then those investment dollars were in a state with return −1%. Going through all securities, we will have their distribution across all possible states such as shown in FIGURE 4.1. Starting with the initial 0% return the investment dollars spread out to all the possible states where they can end up after some time.

---

[28] (L.D. Landau, 1980)

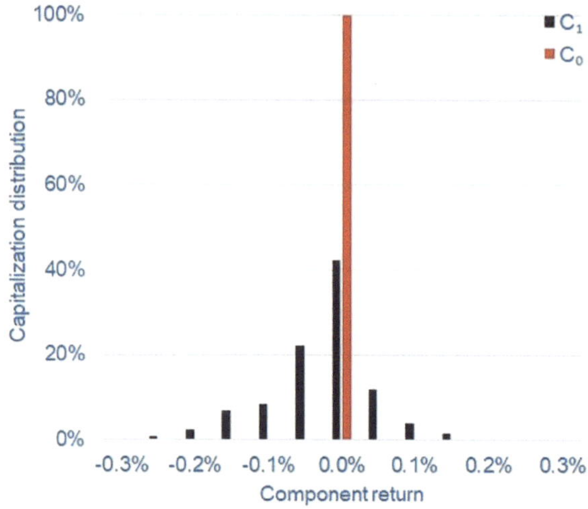

FIGURE 4.1 *Market capitalization by return across different market components (securities). All securities start at* $r_i = 0\%$ *point (orange bar). In the final state prices disperse and return buckets are populated each according to their probabilities (dark bars).*

Macro-ensembles

To make up an ensemble its elements must have common characteristics. Examples of market ensembles are:

- Asset classes: Equities, Bonds, etc.
- Market sectors: Retail, Oil, Biotech, etc.
- Local markets: US, UK, Europe, Emerging markets, etc.

These are the large-scale **macro-ensembles** that are made of securities (or currencies), as opposed to micro-ensembles that are ensembles on micro level.

Micro-ensembles

If there is a market for an individual security, then that market itself can be an ensemble. These are the **micro-ensembles** made of microstructural elements, such as trading orders or price quotes. An order book from an exchange is an example of a micro-ensemble, FIGURE 4.2.

Ask size	Price	Bid size
100	28.20	
900	28.15	
100	27.95	
300	27.90	
100	27.87	
	27.83	100
	27.82	400
	27.80	100
	27.79	1200
	27.78	100
	27.75	300

FIGURE 4.2 *Order book represents a microstructural ensemble of market participants willing to buy and sell an individual security.*

Ensemble idealization

Notion of statistical ensembles usually implies closed systems. This means that they exist and evolve self-sufficiently without impacting the outside world, which gives us the ability to study them separately from the outside world. Financial markets are different: they are the financial world; they are interconnected and are never at equilibrium (in the intuitive sense). On top of that, companies merge, default, securities get delisted, restructured, etc. Strictly speaking, financial markets are open non-equilibrium systems and that is why the ensemble view is an idealization. Despite that, as we will see, a lot of useful information can be derived about them with the ensemble perspective.

4.3 Ensemble measures of financial markets

4.3.1 Market and volatility indices

The most trivial and common ensemble measure of the markets is the weighted **market average index**, which is defined as:

$$S = \sum_{i=1}^{N} n_i \, s_i \tag{4.1}$$

Here n_i is the number of securities, included in the index. Numbers n_i are chosen to reflect capitalization usually as the number of shares outstanding or as the free-float. Examples of market indices are the Standard and Poor's S&P 500 index and the NASDAQ Composite.

Index returns can be expressed through returns of its components as:

$$R = \sum_{i=1}^{N} w_i r_i \tag{4.2}$$

where $w_i = \frac{n_i s_i}{S}$ are the weights of securities in the index. Therefore, index return is just an ensemble-average of individual returns.

Consider a market in which securities have different returns r_i but same volatility:

$$r_i = \bar{r} + \xi_i \tag{4.3}$$

where $\xi_i = \sigma z_i$ and z_i are normally distributed and uncorrelated. Then starting initially with $r_i = 0$ after one step of evolution index capitalization will spread across different r_i as shown in FIGURE 4.1. Width of this distribution will be equal to σ. Therefore, measuring that width will provide volatility of individual securities.

Index volatility can be expressed through volatility of individual securities. Indeed, index return is equal

$$R = \bar{r} + \sigma \sum_i w_i dz_i \tag{4.4}$$

and its variance is:

$$\Omega^2 = \langle (R - \langle R \rangle_T)^2 \rangle_T = \sigma^2 \sum_i w_i^2 \tag{4.5}$$

Therefore, σ can be measured from volatility of the index:

$$\sigma^2 = \frac{\Omega^2}{\sum_i w_i^2} \tag{4.6}$$

If individual volatilities are unequal, then:

$$\Omega^2 = \sum_i w_i^2 \langle \xi_i^2 \rangle_T = \sum_i w_i^2 \sigma_i^2 \tag{4.7}$$

where σ_i is volatility of an individual security, i.e., market variance is the ensemble average of variances with quadratic weights.

Technically, Ω^2 can be represented as either time or ensemble average:

$$\Omega^2 = \langle\langle\xi^2\rangle_T\rangle_{e,w^2} = \langle\langle\xi^2\rangle_{e,w^2}\rangle_T \tag{4.8}$$

where $\langle\ldots\rangle_T$ stands for averaging over timeframe T and $\langle\ldots\rangle_{e,w^2}$ indicates ensemble averaging with quadratic weights. From here we see that

$$\Omega_e^2 = \langle\xi^2\rangle_{e,w^2} = \sum_i w_i^2 \xi_i^2 \tag{4.9}$$

is the ensemble variance of the market, whose time average is equal to historical variance:

$$\langle\Omega_e^2\rangle_T = \Omega^2 \tag{4.10}$$

Based on Eq. (4.6) it is plausible to define an **Ensemble Volatility Index** σ_e (the e-VOL index) such that

$$\Omega_e^2 = \sigma_e^2 \sum_i w_i^2$$

representing the ensemble volatility of individual securities (Sarkissian, 2016):

$$\sigma_e = \sqrt{\frac{\sum_i w_i^2 \xi_i^2}{\sum_i w_i^2}} \tag{4.11}$$

Just like historical volatility, the σ_e index must specify the reference time interval, on which it is based. For example, index based on 1-minute returns must reference 1 minute. It must also reference the underlying market or index that went into its calculation.

The σ_e index represents the current ensemble average volatility of market components as opposed to their historical volatility, based on lengthy time series, or implied volatility, which reflects future expectations. Its physical meaning is that within 1-minute interval individual returns in the market spread around their average value $\bar{r}$ with a distribution that has standard deviation σ_e.

4.3.2 Correlation index

Similar concept can be used to calculate correlations (J. Sarkissian, 2018). The only difference is that correlations are not weighted and therefore the weight factor must be omitted.

Consider a market of securities with the following returns:

$$r_i = \sigma\sqrt{\rho}\, dZ + \sigma\sqrt{1-\rho}\, dz_i \tag{4.12}$$

Here dZ and dz_i are normally distributed random variables, of which the dZ corresponds to a common factor and dz_i corresponds to a specific factor. Time average pairwise correlations of these returns are:

$$\gamma_{ij} = \frac{< r_i r_j >_T}{\sigma^2} = \rho + (1-\rho)\delta_{ij} \tag{4.13}$$

In other words, $\gamma_{i\neq j} = \rho$ and $\gamma_{ii} = 1$, showing that any two securities have correlation ρ and any security is perfectly correlated with itself. Each security has correlation ρ with the index:

$$\frac{< r_i R >}{\sigma^2} = \rho + (1-\rho)w_i \approx \rho \tag{4.14}$$

since $w_i \sim \frac{1}{N}$. Standard deviation of index's returns is $\sqrt{\rho}$ times smaller than that of individual securities:

$$\frac{< R^2 >}{\sigma^2} = \rho + (1-\rho)\sum_{i=1}^{N} w_i^2 \approx \rho \tag{4.15}$$

Note, that this relation does not depend on the number of index components N as long as N is large. Proximity between returns of DJI index with 30 components and returns of S&P 500 with 500 components is an example of that.

Using Eq. (4.13) average market covariance can be written as

$$\langle < r_i r_j >_T \rangle_{i \neq j} = \rho\sigma^2 \tag{4.16}$$

where $\langle \dots \rangle_{i \neq j}$ represents pairwise averaging. After switching the pairwise averaging with time averaging, we get:

$$\langle < r_i r_j >_{i \neq j} \rangle_T = \rho \, \langle < r_i^2 >_i \rangle_T \tag{4.17}$$

Finally, the corresponding **Ensemble Correlation Index** (*e-CORR*) can be defined as:

$$\rho_e = \frac{\langle r_i r_j \rangle_{i \neq j}}{\langle r_i^2 \rangle_i} = \frac{1}{N-1} \frac{\sum_{\substack{i,j=1 \\ i \neq j}}^{N,N} r_i r_j}{\sum_{i=1}^{N} r_i^2} \tag{4.18}$$

This definition makes sure that in this model the time average of covariance based on ensemble averaging and the ensemble average of covariance based on time averaging are equal. Just like σ_e, the ρ_e index must reference the time interval and the underlying market.

If markets had infinite number of securities, values of ρ_e would fall strictly within the range between 0 and 1. For real markets with limited number of securities ρ_e can occasionally take slightly negative values, which essentially means that it is zero.

4.3.3 Relation between volatility and correlation

For a single time step Eq. (4.12) reads:

$$r_i = \bar{r} + \sigma_e \, dz_i \tag{4.19}$$

where $\bar{r} = \sigma\sqrt{\rho}\, dZ$ is the common factor realized over the time step under consideration and $\sigma_e = \sqrt{1-\rho}\,\sigma$. Common factor $\bar{r}$ approximately equals to index return: $\bar{r} \approx R$.

Computing the ρ_e index with this model gives

$$\rho_e \approx \frac{R^2}{R^2 + \sigma_e^2} \tag{4.20}$$

As such the ρ_e index measures the fraction of ensemble variance attributed to common factor out of its total value.

4.3.4 Market drawdown

An important measure associated with market index is the **market drawdown**:

$$D(t) = -\left[\frac{S(t)}{\max\left(S(t' \le t)\right)} - 1, \quad 0\right] \tag{4.21}$$

It shows the percentage by which market index has fallen from its top value in the studied time interval. Since drawdown depends on choice of time interval, it is not a universal measure of the market. Despite that, this measure has gained an important place in finance due to the fact that large drawdowns are firmly associated with financial crises and numerous unpleasant situations accompanying them.

Large drawdowns occur when the market experiences a sudden negative reevaluation. For the most part, the exact size of reevaluation is initially unknown. As market participants try to come to consensus the drawdown accumulates until the market reaches the new level. In that sense drawdown represents an approximate amount of reevaluation (or repricing) occurring to the market. Reevaluation shift is the primary quantity, but since it is unknown, drawdown serves as a proxy to it.

4.4 What is wrong with traditional measures of volatility

Commonly two types of volatility are used: historical volatility (*HV*), which is derived from time series of past prices and implied volatility (*IV*), which is derived from the market price of a traded derivative, particularly options.

Historical volatility is given by Eq. (1.4) and reflects the realized or past volatility. In equity markets implied volatility *IV* is defined as the volatility, which equalizes prices produced by the Black-Scholes option pricing model to market prices:

$$IV\colon\ V\left(s, K, T, \mathbf{IV}, r_f\right) = P \tag{4.22}$$

where s, K, T, r_f and P are the stock price, strike price, time to expiration, risk-free rate, and option market price. Implied volatility reflects future volatility expected over time T. Both *HV* and *IV* have established themselves as widely used volatility measures.

Neither of the two volatility measures describes the current, or instantaneous, volatility. In the meantime, it is mostly the latter that concerns traders, asset managers and risk managers. Due to various reasons[29], HV and IV and their variations are used as a proxy for immediate volatility. Same is true about historical correlations, HC. Let us looks at some of the deficiencies that these quantities have as proxies.

Time lag

Both historical and implied volatilities relate to extended time intervals. Historical volatility uses series of ticks of backward data, which delays its responsiveness. Implied volatility is based on the time until the expiration of the derivative, and as such represents only an average sentiment over that period. Ensemble measures rely on current observables and do not suffer from these deficiencies.

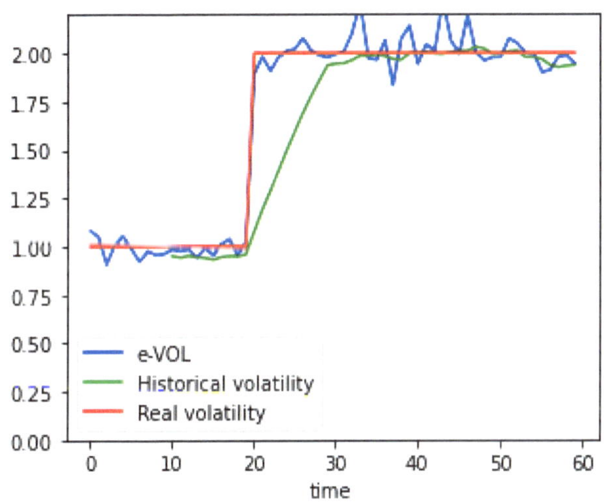

FIGURE 4.3 *Historical volatility and ensemble volatility versus real volatility. HV takes time to adjust to new level and σ_e responds immediately.*

---

[29] These reasons boil down to outright simplicity.

Difference of performance between historical and ensemble volatilities is shown in FIGURE 4.3. The chart shows results of market model with manually set volatility turning abruptly from 1 to 2 in a step-function. Ensemble volatility responds immediately, while historical volatility takes time to adjust. Same is true about historical correlation. In FIGURE 4.4 the real correlation increases in a step function from 0 to 1, but historical correlation needs time to clear the old data from its history. Time lag is a serious deficiency of historical market measures.

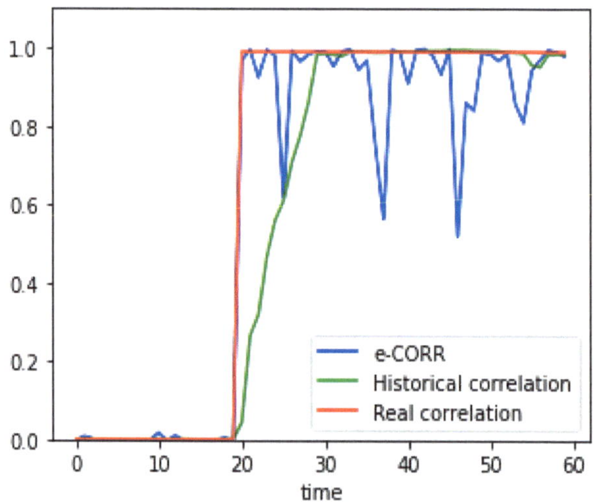

FIGURE 4.4 *Historical and ensemble correlation versus real correlation. Difference of response times is apparent.*

Redundancy of VIX

The role of *IV* index in US equity markets is played by the CBOE Volatility Index (VIX). VIX index has stable anticorrelation with S&P 500, which is well-known among traders. This feature is shown in FIGURE 4.5. There are various views why this happens, and the most accepted one is that traders are afraid that a market drop may develop into a crash whilst index increase is unlikely to turn into an abrupt rally. Whatever the reasons, VIX ends up repeating the inverted moves of S&P 500 and there is a redundancy in such behavior. Due to these factors, it is impossible to say that VIX represents an objective measure of current S&P 500 volatility.

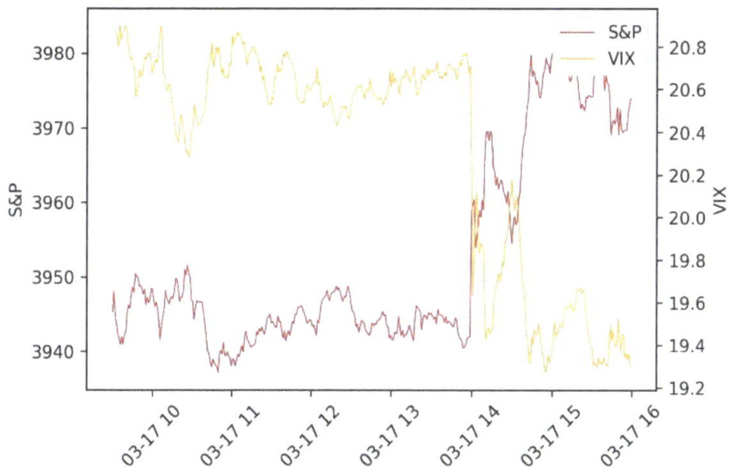

FIGURE 4.5 *VIX and S&P 500. VIX mostly repeats the reversed movements of the underlying index.*

Extended memory

Volatility spikes after sharp market moves are natural: they are needed to facilitate larger jumps. After a jump occurs and the market returns to smaller moves, volatility must return to its previous level. This behavior of σ_e can be traced in FIGURE 4.6. As the market rallies at 2:00 pm, σ_e spikes and afterwards returns to normal levels once the market goes on with smaller moves. At the same time, instead of spiking, VIX drops and stays at the new level even though sharp moves had stopped, FIGURE 4.5. VIX's return to pre-event levels can only be warranted by short-term return of the market to its previous state, or by a substantial time lapse.

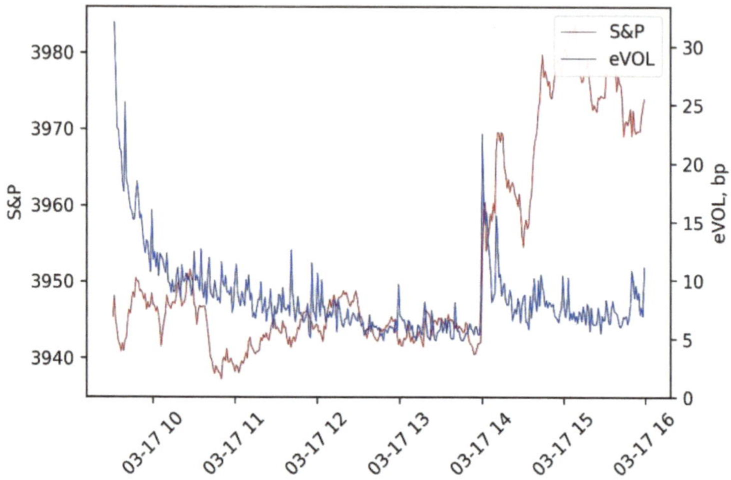

FIGURE 4.6 *Ensemble volatility and S&P 500.*

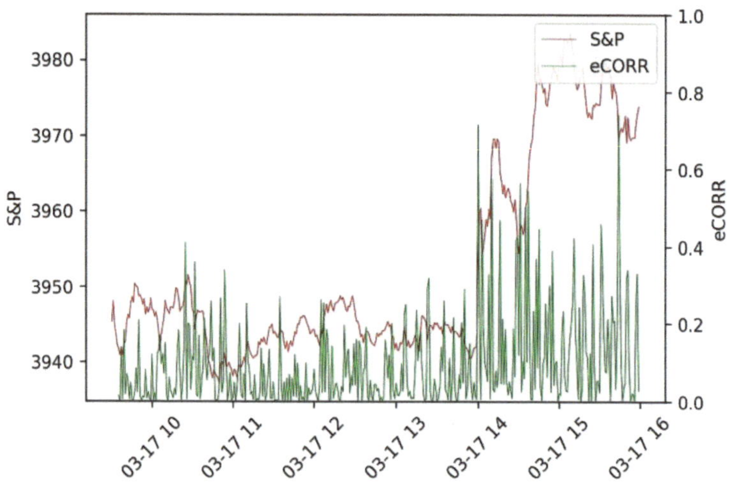

FIGURE 4.7 *Ensemble correlation and S&P 500.*

Certainly, memory is a natural attribute of volatility. But as long as volatility is meant to relate to the size of price moves, it has to reflect that size, and not traders' sentiment. Delayed and inadequate relaxation of VIX is an artefact of its overextended memory which is not a natural quality of volatility.

4.5 Ensemble measures from quantum framework

One of the merits of the coupled-wave model is that it allows to map a security onto a 2-level quantum system. Consequently, a market of securities can be modeled as an ensemble of quantum chaotic elements each of which is represented by a return operator with identical statistics.

Ensemble volatility can be calculated for return operator from Eq. (2.19):

$$\sigma_e = \sqrt{\langle\psi|\boldsymbol{R}^2|\psi\rangle - \langle\psi|\boldsymbol{R}|\psi\rangle^2} = \sqrt{\sigma^2 + \left(\frac{\epsilon}{2}\right)^2 (1 - I^2)} \tag{4.23}$$

Similar calculation can be performed for different market conditions by using an appropriate return operator.

For correlation we have:

$$\rho_e = \frac{1}{N-1}\frac{\sum_{i\neq j} r_i r_j}{\sum_i r_i^2} - \frac{1}{N-1}\frac{\sum_{i\neq j}\langle i|\boldsymbol{R}|i\rangle\langle j|\boldsymbol{R}|j\rangle}{\sum_i\langle i|\boldsymbol{R}^2|i\rangle} = \\ = \frac{1}{N(N-1)}\frac{\sum_{i\neq j}\left(z_i + \frac{\epsilon}{2}I_i\right)\left(z_j + \frac{\epsilon}{2}I_j\right)}{\sigma^2 + \frac{\epsilon^2}{4}} \tag{4.24}$$

With quadratic accuracy in imbalance, this can be written as

$$\rho_e \approx \frac{\left(\frac{\epsilon}{2}\right)^2}{\sigma^2 + \left(\frac{\epsilon}{2}\right)^2} \langle I \rangle^2 \tag{4.25}$$

where ϵ is the value of spread in balanced state and $\langle I \rangle^2 = \frac{1}{N(N-1)} \sum_{i \neq j} I_i I_j$ represents the average quadratic imbalance. This result shows that up to a coefficient ensemble correlation measures the average quadratic imbalance in the market. Thus, market correlations allow to infer the behavior of imbalance.

With a few manipulations we can show that

$$\rho_e = \frac{\left(\frac{\epsilon}{2}\right)^2 \langle I \rangle^2}{\sigma_e^2 + \left(\frac{\epsilon}{2}\right)^2 \langle I \rangle^2} \tag{4.26}$$

which matches Eq. (4.20) with $R = \frac{\epsilon}{2} \langle I \rangle$.

4.6 How to interpret ensemble measures in trading

Reaction speed and meaningfulness of the ensemble measures define their primary uses. As a fast measure of price variations, the most obvious application of σ_e is to serve as a timely signal of variations in volatility and spread levels. The σ_e index also plays a role of liquidity cost measure for market takers and liquidity premia measure for market makers.

During spikes of ensemble volatility, it is possible to lower transaction costs by applying larger price offsets to limit orders and then lowering them as σ_e drops. On the risk side, high σ_e levels signal the possibility of haircut revision and the need for increased margin requirements.

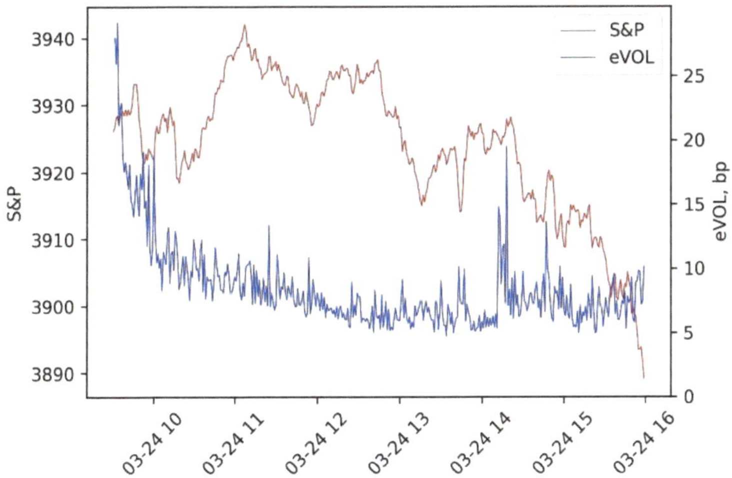

FIGURE 4.8 *S&P index and* σ_e

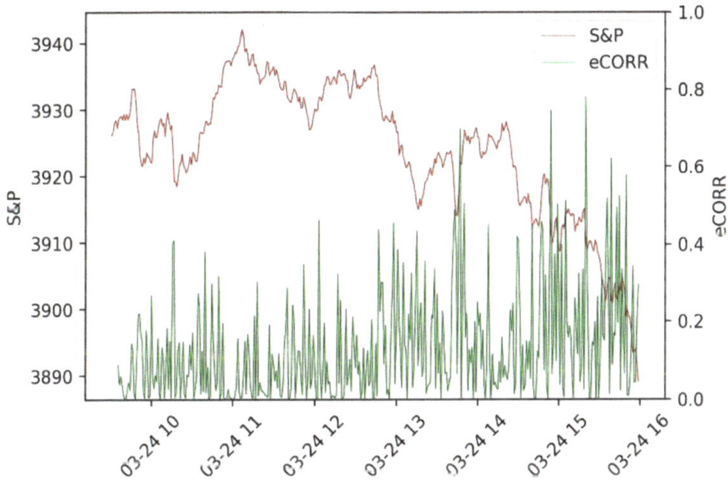

FIGURE 4.9 *S&P index and* ρ_e

The role of ρ_e is in helping to characterize market state without considering economic reasons. In Section 4.5 we saw that ρ_e

measures average quadratic imbalance. As such, large ρ_e indicates a highly imbalanced market usually accompanied by a continuing directional movement. Values $\rho_e > 70\%$ accompany market crashes or rebounds. Low ρ_e indicates either a market with balanced securities or a market with imbalanced but disorganized securities[30].

FIGURE 4.8 and FIGURE 4.9 show σ_e and ρ_e on a common day. Ensemble volatility starts the day at 25 bp and then gradually decreases to its intraday level of about 5-6 bp. Opening σ_e is always large capturing the fact that the markets open with high volatility and large spread. This is not a critical behavior (H. Kumar, 2019). It has to do with the overnight trading halt, during which price uncertainty accumulates since there are no transactions.

Ensemble correlation often opens at zero level. Even though opening imbalances of individual securities may be large, they are usually disorganized as traders try to establish new price levels, so their ensemble average is very small.

After about an hour of trading σ_e drops to 7-8 bp. Market correlation ρ_e raises to 10-30% levels, which is also a normal level and tells us that securities interact in a normal mode. Slight increase of σ_e at the end of the day is also usual. Despite the fact that bid-ask spread tends to decrease in the last minutes of trading, the high-low range increases, and that increase is captured by σ_e.

Usually, prior to a market crash the level of σ_e is already elevated compared to the adjacent days. When the crash begins, σ_e spikes and ρ_e quickly raises to >60% as seen in FIGURE 4.11 and FIGURE 4.12. Top of the spike is reached simultaneously with market's local bottom, FIGURE 4.13.

---

[30] Such as observed at the opening.

After the drawdown, σ_e drops even if the market did not climb out of the drawdown. Correlations, on the other hand, continue to stay large for a longer time. Only when correlations go down can we diagnose the market as going back to the normal state.

4.7 Measuring market ergodicity

Ensemble volatility and correlation indices σ_e and ρ_e are equal to their time average counterparts if all securities are completely random and independent. This is of course an idealistic case which does not occur in the markets. As a result, time and ensemble averages diverge. The amount of divergence is determined by market ergodicity[31]. Question arises: how can market ergodicity be measured?

Market ergodicity means that securities can attain any return from the spectrum of **the same** return operator. In other words, if we took a simultaneous measurement of returns of all securities, their distribution would be an identical fit with the spectrum of the return operator. If some region of returns stands out and is visited frequently while other regions are avoided, ergodicity breaks. For instance, in coupled-wave model the return operator has two eigenstates. If securities keep preferring the bid price while avoiding the ask price, that is a non-ergodic behavior.

Non-ergodic behavior is typical of correlated markets. The ensemble correlation index ρ_e can therefore serve as a measure of market ergodicity. In ergodic markets $\rho_e \approx 0$ and in non-ergodic markets $\rho_e \approx 1$.

---

[31] (G.G. Malinetsky, 2000)

So, how correlated are the markets? Let us take US equity market as an example. One way that correlations present themselves is that index volatility is substantially larger than it should be in an uncorrelated market. For instance, an average DJI component stock has daily volatility around 1.7%. Given that there are 30 components in DJI, index volatility would have been 0.3% if its components were uncorrelated. That number would have been even lower for S&P 500, which has 500 components. In the meantime, average DJI and S&P 500 volatilities are approximately the same and are around 1.1%. From here we can deduce that the implied market correlation is about 0.42.

$\langle r \rangle \approx 0$ $\langle r \rangle > 0$

FIGURE 4.10 *Disordered and ordered states of the market. Individual returns are shown with arrows. Their geometric positions on the chart are for illustrational purposes only and have no meaning.*

Ergodicity breaking is deeply connected to transitions between order and disorder (L.D. Landau, 1980). The reason that individual states fill up the entire return spectrum is that they are random. In an ordered market securities cluster around only a subset of states. Ordering direction is determined by market conditions. It removes the symmetry between $+r$ and $-r$ that exists in a disordered state. If returns were schematically put on an axis, the difference between two states could be represented by FIGURE 4.10.

The extent of ordering is described by the **order parameter**:

$$\eta = \frac{1}{N(N-1)} \sum_{\substack{i,j=1 \\ i \neq j}}^{N} I_i I_j \tag{4.27}$$

Its value is zero in a disordered state and non-zero in an ordered state[32]. We have seen this quantity to come up in Eq.(4.25) for the correlation index. Both the average quadratic imbalance and correlations can be used as measures of ergodicity or non-ergodicity. Ordered states can be easily recognized as crashing or rallying markets, depending on the direction.

4.8 Second-order phase transitions in financial markets

Among all phenomena in financial markets, market crashes, price corrections and market rallies present special interest. These events share one common feature: they involve large price moves within short period of time. They are known to be responsible for so many unexpected trading losses, margin calls and defaults. This is why these events attract special attention. Their understanding is crucial for decision making in asset management, trading, and generally for market stabilization. Here these events – market crashes, price corrections, rallies, and rebounds – will be collectively referred to as **critical market events**.

In previous section we saw that an ordered market state corresponds to strongly directional market and a disordered state corresponds to a market without direction (Sarkissian, 2019). These

---

[32] (L.D. Landau, 1980)

states are more appropriately referred to as market phases, reserving the notion of "state" for characterization with average imbalance $\langle I \rangle$. Critical events occur as transitions between the phases.

Order-disorder transitions are generally known to be second-order type. Indeed, markets do not exhibit phase coexistence or latent transfer effects, so the possibility of transitions being the first-order type is highly unlikely. Second-order phase transitions are characterized by the following features (L.D. Landau, 1980):

1. There must be distinct separate phases of the ensemble
2. Phases demonstrate different organizational order
3. There must be an order parameter that takes different values in different phases
4. Correlations must raise near the transition point between two phases
5. Fluctuations must spike near the transition point
6. Switching from one phase to another and back must show a hysteresis in the order parameter.

These behaviors must be based on observable and measurable input. Relying on notions like "investor attitude", "market sentiment" and "market temperature" that cannot be measured is insufficient. The ensemble measures provide a solid basis that makes it possible.

Unlike common physical systems, where temperature or external field can be set in an experiment, no one can fix the parameters of financial markets. Instead, they continuously evolve while adjusting to new conditions. This will show as a trajectory on the phase diagram. We cannot expect the market to form a specific phase and stay there.

Although only crashes will be described here, these are the ones presenting the most interest. Mechanisms governing the dynamics of rallies are similar, but less pronounced.

4.8.1 Ensemble measures in critical events

Typical critical behavior of ensemble measures is shown in FIGURE 4.11- FIGURE 4.13. These are 1-minute ensemble measures for S&P 500 index on Aug. 1, 2019 when the index went into a substantial intraday drawdown (>1%). More charts are given in APPENDICES A-D. On that date the Fed announcement came in at 1:30 pm.

As explained before, large volatility in the beginning of the day is normal. This is not a critical behavior and has to do with the overnight trading halt, during which price uncertainty accumulates.

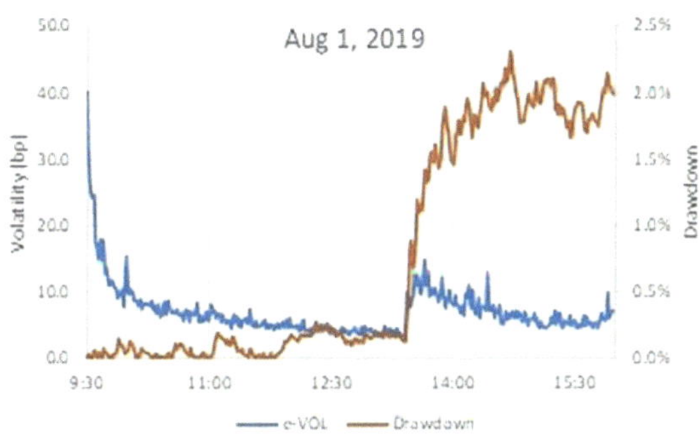

FIGURE 4.11 *Impact of drawdown on 1-minute σ_e for S&P 500 index.*

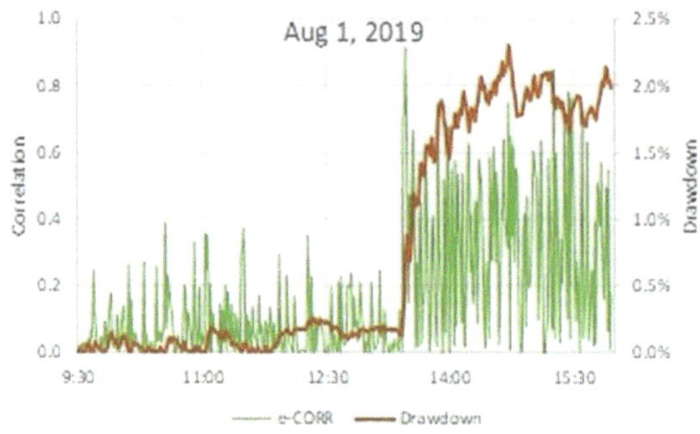

FIGURE 4.12 *Impact of drawdown on 1-minute ρ_e for S&P 500 index.*

FIGURE 4.13 *S&P 500 index and its drawdown.*

Market drawdown is accompanied by a sharp volatility spike and correlation increase. Both react to drawdown immediately. Volatility and correlation have similar response time, but different relaxation times. After the impact, correlations persist longer than volatility, which drops down from the peak soon after the critical event.

These effects are only observed for fast and large drawdowns. A gradually accumulating drawdown does not initiate volatility peaks or large correlations.

4.8.2 Order parameter and market phases

Ensemble phases are differentiated by the order parameter. Therefore, it makes sense to look at the average value of ensemble correlation ρ_e, playing the role of order parameter, across different ranges of drawdown and volatility. This is shown in FIGURE 4.14 and FIGURE 4.15. Cell color reflects the value of order parameter in the cell: from deep red for 0 (no correlation), through yellow (some correlation), to deep green for 1 (perfectly correlated).

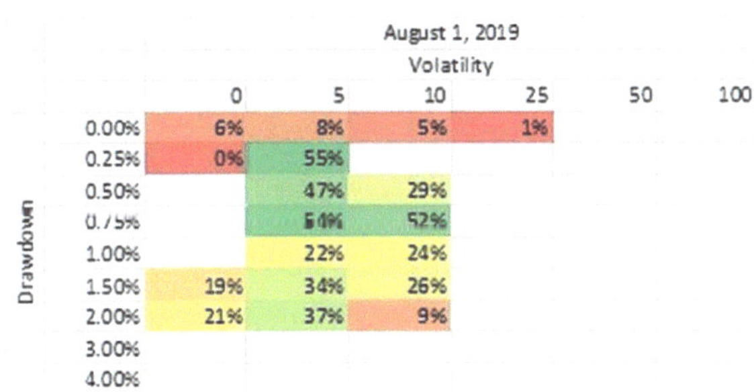

FIGURE 4.14 *Correlation levels vs. drawdown and 1-minute ensemble volatility for S&P 500 index on a day with substantial impact. Volatility is given in basis points.*

June 18, 2019

Drawdown	Volatility 0	5	10	25	50	100
0.00%	8%	9%	12%		50%	
0.25%	10%	16%				
0.50%						
0.75%						
1.00%						
1.50%						
2.00%						
3.00%						
4.00%						

FIGURE 4.15 *Correlation levels vs. drawdown and 1-minute ensemble volatility for S&P 500 index on a day without substantial impact. Volatility is given in basis points.*

March, 2008 - December 2010

Drawdown	Volatility 0%	1%	2%	3%	5%	15%
0.00%	1%	15%	32%	51%		
5.00%		8%	62%	88%		
10.00%	8%	19%	46%	53%	27%	
15.00%	12%	29%	44%	58%	60%	
20.00%	5%	30%	62%	80%	75%	
25.00%	5%	20%	45%	38%	74%	
30.00%	19%	10%	22%	45%	68%	
40.00%		1%	31%	35%	67%	
50.00%						

FIGURE 4.16 *Correlation levels vs. drawdown and 1-day ensemble volatility for Dow Jones Industrial Average index in 2008-2010.*

Similar diagram on a larger time scale is shown in FIGURE 4.16 which shows correlation levels for the Dow Jones Industrial Average index based on 1-day returns as it goes through the 2008-2010 crisis.

The two market phases can be visually identified as red-orange for uncorrelated phase and green-yellow for strongly correlated phase. The schematic phase diagram is shown in FIGURE 4.17. Remembering that correlation measures average quadratic imbalance, the two phases can also be characterized as balanced and strongly imbalanced. The intuitive labeling of the phases would be "**normal phase**" and "**critical phase**".

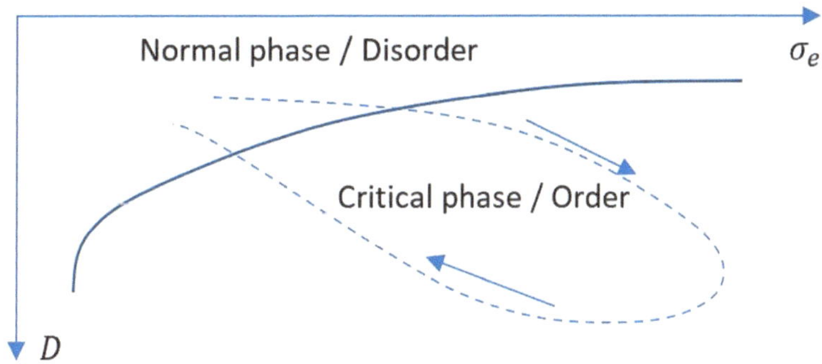

FIGURE 4.17 *Schematic representation of phase diagram*

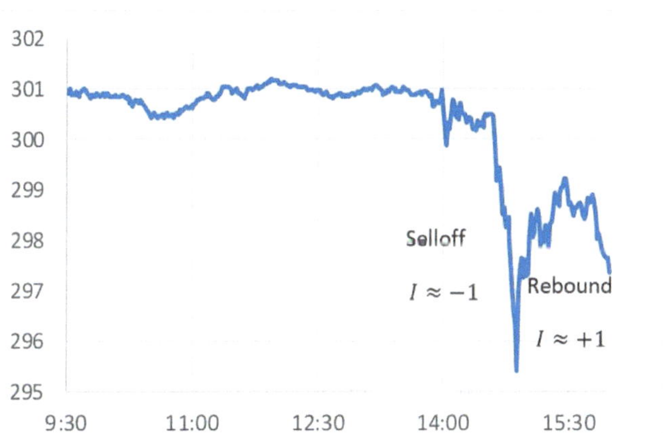

FIGURE 4.18 *S&P 500 tracking ETF SPY on July 31, 2019. State with $I \approx -1$ changes to $I \approx +1$ without going through the balanced phase.*

Zero imbalance $\langle I \rangle = 0$ is the equilibrium value in normal phase. Any buildup of imbalance will have tendency for reduction through gradual price adjustment in the direction of imbalance.

Due to quadratic dependence of ρ_e on $\langle I \rangle$ correlation 0.6-0.8 translates into imbalance of 0.8-0.9. Both states with strongly positive and negative imbalance ($\langle I \rangle \approx +1$ and $\langle I \rangle \approx -1$) fall into critical phase. Zero imbalance $\langle I \rangle = 0$ is not an equilibrium state in a critical phase. Switch between $\langle I \rangle \approx -1$ and $\langle I \rangle \approx +1$ occurs abruptly without an intermediate passage through the normal phase, as can be seen in FIGURE 4.18. We saw in Section 2.13 how this switch occurs dynamically due to level coupling. But from statistical perspective that involves many transactions per unit of time it appears abrupt.

4.8.3 Critical hysteresis

As the market goes through a critical event its path on volatility-drawdown diagram draws a loop as shown in FIGURE 4.17. The loop shows that the system returns to the normal state in a path different from the one that took it out. This behavior, known as the hysteresis, is typical of ergodicity-breaking dynamics.

Technically, the loop is a result of asymmetry between market response and its subsequent relaxation. This asymmetry can be captured and measured by the hysteresis function, defined as

$$H(t_n) = \sum_{t_i=0}^{t_n} \sigma_e(t_i)\,[D(t_i) - D(t_{i-1})] \tag{4.28}$$

and representing the area under $\sigma_e[D(t)]$ curve. This area accumulates with each loop. As a result, $H(t)$ grows after each volatility spike as shown in FIGURE 4.19.

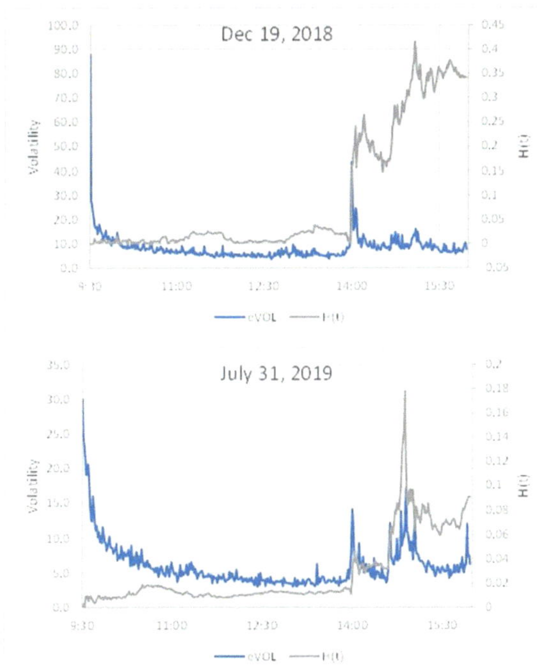

FIGURE 4.19 *Hysteresis function H(t) raises after volatility spikes due to volatility-drawdown hysteresis.*

4.8.4 Bifurcation of imbalance

The main cause of the phase transition is sudden market repricing, which is represented by the drawdown. Phase transition is observed only for large price shift and does not occur for a small shift. If the amount of repricing is small, price adjusts gradually through random walk without rising correlations and volatility.

Order parameter is obviously a function of drawdown and volatility. Although volatility is also caused by drawdown, it can be factored out. We can therefore write:

$$\rho_e = f[D, \sigma(D)] \tag{4.29}$$

were f is some function typical for a phase transition at $D > D_c$ schematically shown in FIGURE 4.20

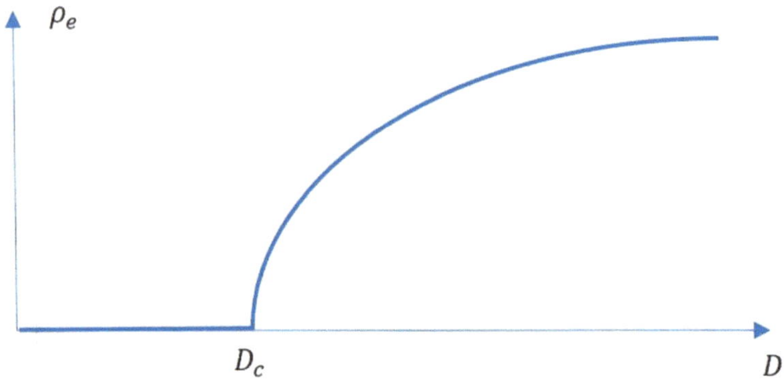

FIGURE 4.20 *Order parameter buildup for large drawdown $D > D_c$.*

Converting Eq. (4.28) into a relation for imbalance using Eq. (4.25), we can see that imbalance branches out and exhibits bifurcation[33] for $D > D_c$ as shown in FIGURE 4.21:

$$\langle I \rangle \approx \pm\sqrt{f[D, \sigma(D)]}$$

This is why in critical state all securities in the market can simultaneously flip the sign of their imbalance without continuously going through zero in between. The $\langle I \rangle = 0$ state is not even an available choice. The balanced state has to be reached gradually as the market climbs out of the critical phase.

---

[33] (G.G. Malinetsky, 2000)

As mentioned before, drawdown itself is not a completely objective measure. It is being used it only because it is firmly associated with critical events and market repricing. Unlike the repricing amount, which is unknown at the time of the event, drawdown is measurable at any time. In this sense it serves as a proxy of the repricing shift. If the amount of repricing were known, the imbalance would be a function of price shift, not of drawdown:

$$I = \pm\sqrt{g[\delta s, \sigma(\delta s)]} \tag{4.30}$$

Here $\delta s = s - s_0$ is the repricing amount, $s(t)$ is the current market value, s_0 is the new fair value, and g is another nonnegative function similar in behavior to f. This takes us to the notion of **imbalance field** $I(s)$, which shows what imbalance the security would experience if it were traded at price s at an offset from its fair value. From this section on, imbalance will be used in that sense.

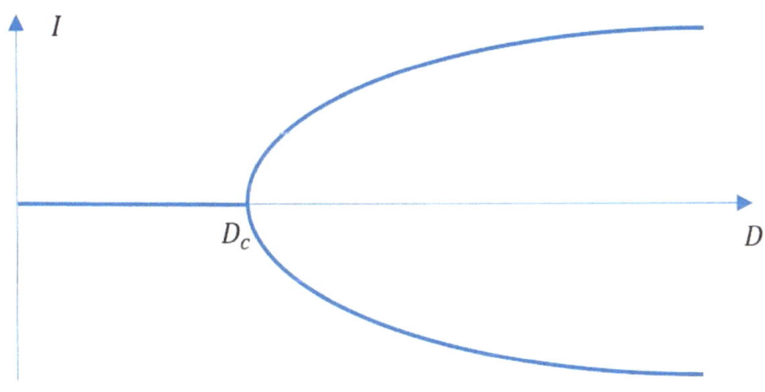

FIGURE 4.21 *Bifurcation of imbalance I for large drawdown D > Dc.*

4.9 Ensemble theory of price stabilization

4.9.1 Fair value and imbalance field

Classical finance theory teaches that price forms as a result of equilibrium between supply and demand. In classical theory this equilibrium takes form of the intersection between supply and demand curves. If one of them moves, the intersection point moves accordingly, leading to an immediate price shift. Since curves intersect at a single point, price in this model is represented by a single number. That number may fluctuate as a result of fluctuations in supply and demand, but it always remains a single number. We know that on microstructural level price is not a single number. Therefore, on microstructural level the classical notion of supply-demand equilibrium needs revision (Orrell, 2020) (Sarkissian, 2020).

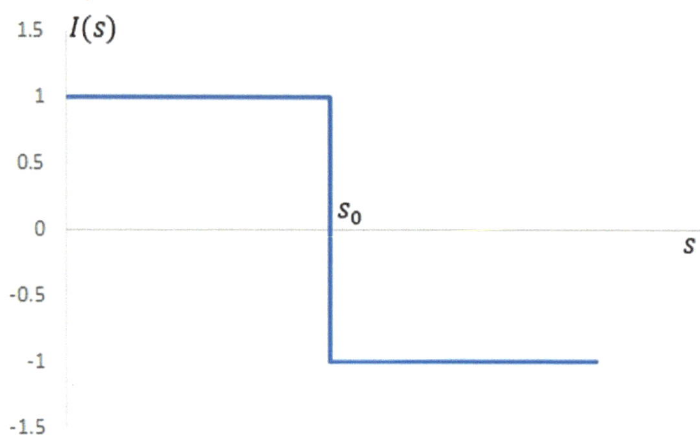

FIGURE 4.22 *Imbalance field for a security with fair price s_0.*

In balanced markets orders must execute at bid and ask levels with equal probability. Therefore, if s_0 is a fair value, then $I(s_0) = 0$. In this sense phrase "fair value" reflects not only the common sense, but also the definitions of IFRS[34] (International Financial Reporting Standards).

If price were a definite quantity, then $I(s)$ would be a step-function as shown in FIGURE 4.22. The meaning of this plot is that if price were definite, then any quantity of security offered at price lower than s_0 would be immediately bought, and any quantity of demand bidding higher than s_0 would be immediately satisfied. Our purpose is to determine $I(s)$ that overcomes the difficulty of classical theory and reflects price uncertainty. To approach that problem, we must begin with the free energy of the market (Sarkissian, 2020).

4.9.2 Market equilibrium equation

Let us compose the free energy of the market as a statistical system. State of the security is completely determined by its imbalance and phase. We have seen in Section 2.13 that near balance phase plays an unimportant role. It can therefore be disregarded, leaving only I in play.

Considering that $|I| \leq 1$, let us write the Landau-Ginzburg expansion of free energy in imbalance and its first derivative:

$$F = F_I + F_S \quad (4.31)$$

[34] According to IFRS the fair value of an asset may be different from its market value. It takes into account not only market conditions, but also other information, particularly risk.

where

$$F_I = a_1 I + \frac{a_2}{2} I^2 + \frac{a_3}{3} I^3 + \frac{a_4}{4} I^4 \tag{4.32}$$

is the imbalance part and

$$F_S = b_1 \frac{dI}{ds} + \frac{b_2}{2} \left(\frac{dI}{ds}\right)^2 \tag{4.33}$$

is the spread part (we will see later what makes this name appropriate). It is easy to check, that even though the cross-terms with product combinations of I and $\frac{\partial I}{\partial s}$ are possible, they will have no effect on the end result. While slight asymmetry between buying and selling generally exists, magnitude of associated effects is such that this asymmetry is unessential. We will therefore drop the odd terms from F_I:

$$a_1 \approx 0,\ a_3 \approx 0$$

$$F_I \approx \frac{a_2}{2} I^2 + \frac{a_4}{4} I^4 \tag{4.34}$$

Summarizing, the total free energy is:

$$F = \frac{a_2}{2} I^2 + \frac{a_4}{4} I^4 + b_1 \frac{dI}{ds} + \frac{b_2}{2} \left(\frac{dI}{ds}\right)^2 \tag{4.35}$$

Interplay between F_I and F_S defines energy minimum and the stable distribution of $I(s)$. To ensure stability in I we must have $a_4 > 0$.

For $a_2 > 0$ energy F_I reaches minimum at $I = 0$. For $a_2 < 0$ the $I = 0$ equilibrium becomes unstable as two other equilibria with lower free energy appear:

$$I = \pm\sqrt{\frac{-a_2}{a_4}} \tag{4.36}$$

Since phase transitions are observed only for large shifts in fair price, this suggests that coefficient $a_2(s)$ remains positive in an interval around the fair price and changes to negative outside of that interval. For example, the following approximation with positive coefficients could be used:

$$a_2 \approx \alpha_0 - \alpha_2(s - s_0)^2 \tag{4.37}$$

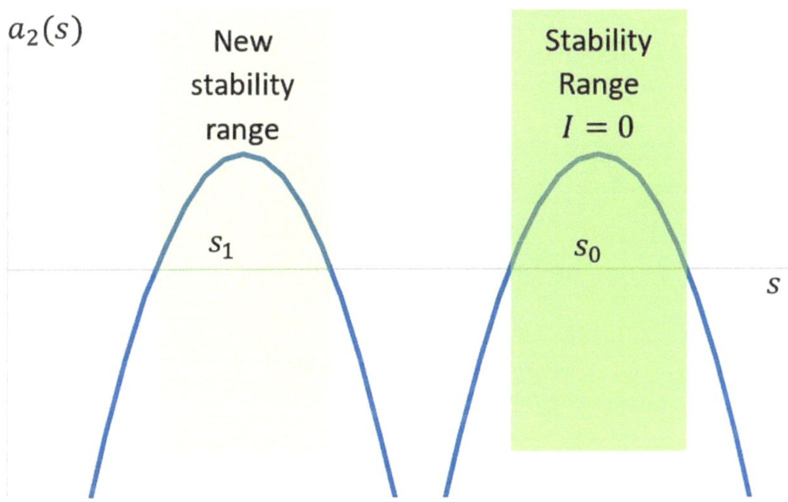

FIGURE 4.23 *Shift in coefficient a_2 before and after repricing from fair price s_0 to s_1*

Naturally, since $|I| \leq 1$ in real systems coefficient a_2 is not going to drop below $-a_4$. The stability region is the range where $a_2(s) > 0$:

$$s_0 - \sqrt{\frac{\alpha_0}{\alpha_2}} < s < s_0 + \sqrt{\frac{\alpha_0}{\alpha_2}}$$

Within this region imbalance will fluctuate around zero and price will move in a random walk, FIGURE 4.23. Outside of this region price will come under a strong pressure from imbalance and move directionally towards the region with $a_2 > 0$.

Since in finance the only physically valid situation is when $\frac{\partial I}{\partial s} \leq 0$ coefficient b_1 should be positive: $b_1 > 0$. Lastly, to ensure stability in $\frac{\partial I}{\partial s}$ coefficient b_2 must be positive $b_2 > 0$.

4.9.3 Fair value and spread

Equilibrium is reached when:

$$\frac{\delta F}{\delta I} - \frac{d}{ds}\frac{\delta F}{\left(\delta \frac{dI}{ds}\right)} = 0 \tag{4.38}$$

Coming from the free energy described by Eq. (4.35), we must therefore have the following equilibrium equation for imbalance:

$$a_2 I + a_4 I^3 - b_2 \frac{d^2 I}{ds^2} = 0 \tag{4.39}$$

In this equation a_2 is a function of s. To begin, we will consider it constant and generalize later.

Apart from the trivial solution $I(s) = 0$, this equation has a few other solutions. If $a_2 < 0$ then

$$I(s) = -I_0 \, tanh\left(\frac{s - s_0}{\Delta s}\right) \quad (4.40)$$

where $I_0 = \sqrt{\frac{|a_2|}{a_4}}$ and $\Delta s = \sqrt{\frac{2b_2}{|a_2|}}$, and the minus sign is chosen to satisfy condition $\frac{\partial I}{\partial s} \leq 0$. Naturally, in real markets we must have $I_0 = 1$:

$$I(s) = -\, tanh\left(\frac{s - s_0}{\Delta s}\right) \quad (4.41)$$

This solution is shown in FIGURE 4.24. For $\Delta s \to 0$ it transforms into the step-function shown in FIGURE 4.22.

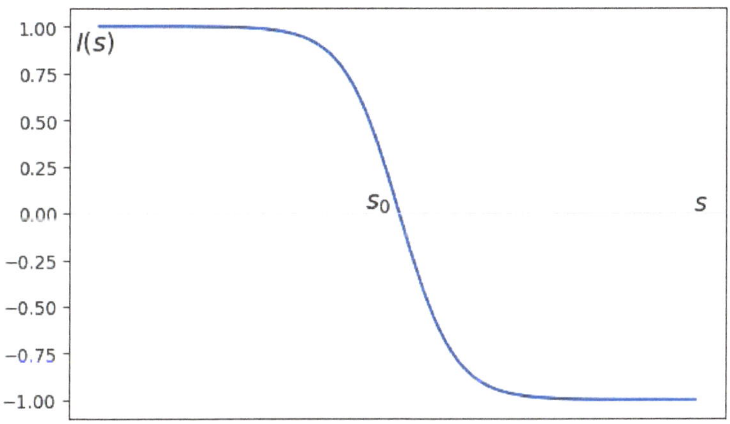

FIGURE 4.24 *Graphical representation of solution Eq. (4.41)*

For $a_2 \geq 0$

$$I(s) = -I_0 \, tan\left(\frac{s - s_0}{\Delta s}\right) \quad (4.42)$$

where $I_0 = \sqrt{\frac{a_2}{a_4}}$ and $\Delta s = \sqrt{\frac{2b_2}{a_2}}$. This solution is periodic and does not satisfy the condition $\frac{dI}{ds} < 0$. It will not realize in real markets. Since the $\tan(x)$ function diverges for arguments close to $\pm\pi/2$ this would lead to formation of numerous stability regions across the price axis with discontinuities between them. As a result, the system will choose the trivial solution $I(s) = 0$ as more energy favorable.

Another, solution of Eq. (4.39) is:

$$I(s) = \frac{I_0}{cosh\left(\frac{s - s_0}{\Delta s}\right)} \tag{4.43}$$

with $I_0 = \pm\sqrt{\frac{-2a_2}{a_4}}$ and $\Delta s = \sqrt{\frac{b_2}{a_2}}$. This solution also violates the condition $\frac{dI}{ds} < 0$.

The only physically valid solution is the one provided by Eq. (4.41). The same result was obtained in Section 3.8 from quantum framework. It is similar to the one presented in FIGURE 4.22 but allows for price uncertainty. Apart from the discrete $I = \pm 1$ format, intermediate values of imbalance are allowed while the main features are preserved: $I(s \ll s_0) = -1$ and $I(s \gg s_0) = 1$, and $I(s_0) = 0$. This is the type of solution that we were looking for.

Note, that solution Eq. (4.41) is valid for any s_0. Remembering that coefficient a_2 is in fact a function of s, it is clear that the form of $a_2(s)$ specifies the position of s_0.

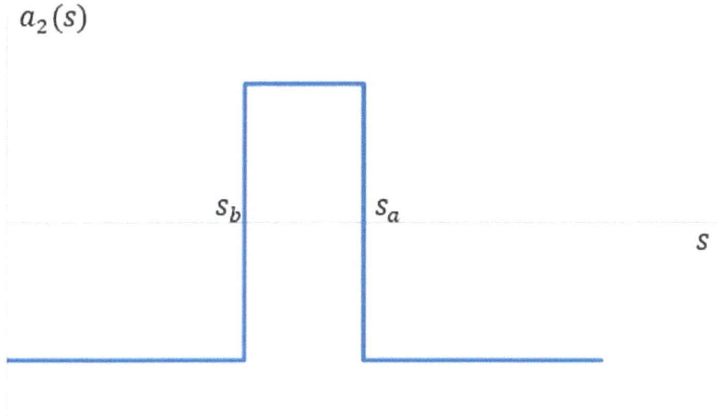

FIGURE 4.25 *Coefficient $a_2(s)$ with rectangular profile*

Consider a model coefficient α_2 with rectangular profile as shown in FIGURE 4.25:

$$\begin{cases} a_2 > 0, & s_b < s < s_a \\ a_2 = -a_4, & s \leq s_b \text{ and } s \geq s_a \end{cases} \tag{4.44}$$

Here s_b represents the bid price and s_a represents the ask price. Strictly speaking, for this profile the equilibrium equation cannot be solved analytically since the derivatives stitching complicates the procedure. However, the stitching contribution to imbalance around the stitching points is less than 0.1, to which the markets are insensitive, and therefore a simple piece-wise solution serves as a good approximation:

$$\begin{cases} I(s) = -\tanh\left(\dfrac{s - s_b}{\Delta s}\right), & s \leq s_b \\ I(s) = -\tanh\left(\dfrac{s - s_a}{\Delta s}\right), & s \geq s_a \\ I(s) = 0, & s_b < s < s_a \end{cases} \tag{4.45}$$

The real solution is shown in FIGURE 4.26. In the limit of $s_b \to s_0$ and $s_a \to s_0$, Eq. (4.45) will converge to Eq. (4.41) positioned at the fair value s_0.

Practically, there is no requirement for interval $s_a - s_b$ to be zero. It can be larger for illiquid securities, allowing for a larger window of price uncertainty. Let us remember that while spreads in equity markets are as thin as a few basis points, they can reach 100's *bp* in much less liquid fixed income and derivatives markets.

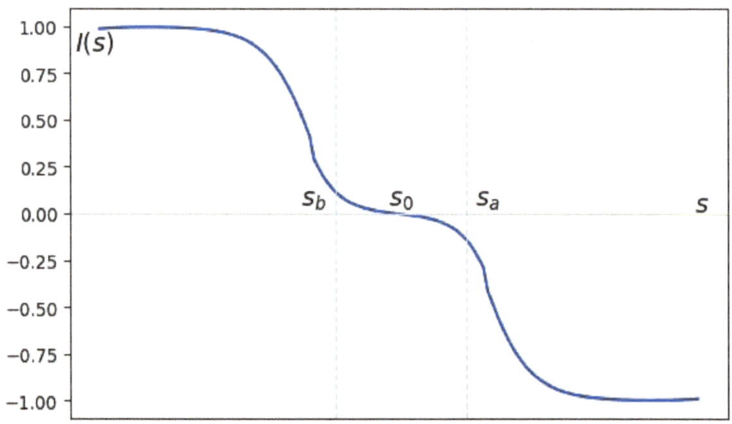

FIGURE 4.26 *Solution of Eq. (4.39) with rectangular profile*

Finally, we can use the Eq. (4.41) to calculate the spread:

$$\epsilon = \frac{\Delta s}{s} = \frac{\sqrt{\frac{2b_2}{a_4}}}{s} \tag{4.46}$$

Presence of spread in FIGURE 4.24 as opposed to FIGURE 4.22 is warranted by the F_s part of the free energy, which is why it is called the spread part.

4.9.4 Price dynamics after fair value shift

Now that we have the functional dependence of imbalance field, we can calculate price dynamics under its influence. Eq. (4.41) describes the steady state of the market. After a shift of fair value, the new equilibrium imbalance field will be centered at the new fair value (Sarkissian, 2020), (Stoikov, 2018). Transition to the new equilibrium happens gradually in a manner coordinated with market price, so $I = I(s, t)$. Consider price dynamics in an imbalance field that has already transitioned, so $I = I(s)$ is centered at a new level. Practically, this corresponds to a situation in which a broad public announcement is made stating the new security price and leaving little space for speculation about it. Examples of such situations are cash acquisitions in equity markets and early redemptions in fixed income. These situations do not result in critical events but are also quite frequent.

Suppose that the fair value shifts from s_0' to s_0 by an amount $\delta s_0 = s_0' - s_0$. The dynamic equation for price under the influence of imbalance follows from the coupled-wave model and is:

$$\tau \frac{ds}{dt} = \frac{\Delta s}{2} I \tag{4.47}$$

which gives the following equation for the remaining deviation $\delta s = s - s_0$:

$$sinh\left(\frac{\delta s(t)}{\Delta s}\right) = e^{-\frac{t}{\tau}} sinh\left(\frac{\delta s_0}{\Delta s}\right) \tag{4.48}$$

If the price shift is large $|\delta s_0| \gg \Delta s$, price converges to the new fair value linearly:

$$\delta s(t) = \delta s_0 - \Delta s \frac{t}{\tau} \tag{4.49}$$

This equation means that with each trade price gets closer to the fair value by the amount of semi-spread[35].

For a small price shift $|\delta s_0| \ll \Delta s$ price converges to the new fair value exponentially and within the characteristic trade time τ:

$$\delta s(t) = e^{-\frac{t}{\tau}}\, \delta s_0 \tag{4.50}$$

Exponential convergence in this case works much slower than linear because $|\delta s_0| \ll \Delta s$. The two regimes can be seen in FIGURE 4.27.

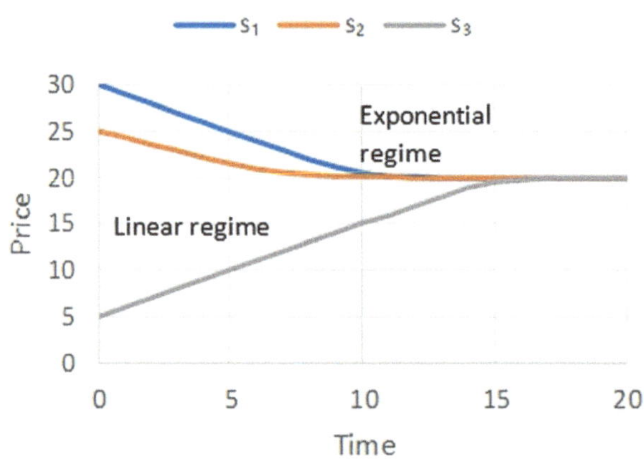

FIGURE 4.27 *Price convergence to* $s_0 = 20$ *for* $s(t)$ *with different parameters* δs_0, Δs *and* τ.

---

[35] From traders' perspective the entire spread is paid in each transaction. In reality only half is paid, and the perception of entire paid spread arises because of the illusion that the other half is also available.

4.10 Order imbalance and execution probabilities

In PART 2 execution probabilities of coupled-wave model were determined from dynamic equations. These equations describe market conditions through matrix elements of the return operator. Another way to determine functional dependencies is from ensemble approach. Here is how it works with an example of execution imbalance.

Consider a market of $K \gg 1$ similar securities with identical statistics, but not with identical current parameters. Generally, different securities in this market will have different order imbalance, and the market state is then determined by the set of order imbalances for each security. Let us compute the number of counts in a time interval to find the market in configuration $N_a(\delta q_1, \delta q_2, \dots, \delta q_K)$ with securities having imbalances δq_i executing at asking price. Since the combination of order books produces the total book, this number is equal $N_a(\delta q_1 + \delta q_2 + \dots + \delta q_K)$ up to a permutation factor Z, and whose particular value is unimportant due to the eventual normalization.

Since securities are independent, N_a must factorize:

$$N_a(\delta q_1, \delta q_2, \dots, \delta q_K) = n_a(\delta q_1) n_a(\delta q_2) \dots n_a(\delta q_K) \qquad (4.51)$$

By taking the logarithm and combining imbalances, we now have

$$ln[N_a(\delta q_1, \delta q_2, \dots, \delta q_K)] = ln[N_a(\delta q_1 + \delta q_2 + \dots + \delta q_K)] + \qquad (4.52)$$

$$+\ln(Z) = ln[n_a(\delta q_1)] + \dots + ln[n_a(\delta q_K)]$$

This means that $ln(n_a) - \frac{\ln(Z)}{K}$ is an additive quantity. The only quantity in the market that is additive in δq is δq itself. Indeed, a combination of two order books produces a book with totals. Therefore, we must have

$$ln\left[\frac{n_a(\delta q)}{\sqrt[K]{Z}}\right] = \frac{\delta q}{C} \tag{4.53}$$

where C is some constant, and so:

$$n_a(\delta q) = \sqrt[K]{Z}e^{\frac{\delta q}{C}} \tag{4.54}$$

Qualitative behavior of $n_a(\delta q)$ warrantees that $C > 0$. Obviously,

$$n_b(\delta q) = n_a(-\delta q) = \sqrt[K]{Z}e^{-\frac{\delta q}{C}} \tag{4.55}$$

From here, the execution probabilities can be found:

$$p_a = \frac{1}{2} + \frac{1}{2}tanh\left(\frac{\delta q}{C}\right)$$

$$p_b = \frac{1}{2} - \frac{1}{2}tanh\left(\frac{\delta q}{C}\right)$$

The execution imbalance is now:

$$I(\delta q) = p_a - p_b = tanh\left(\frac{\delta q}{C}\right) \tag{4.56}$$

Comparing this to Eq. (2.72), for $\left|\frac{\delta q}{C}\right| \ll 1$ we can determine the constant $C = |Q|\frac{T}{\Delta t}$, where $|Q| = m_{ask} + m_{bid}$. Thus,

$$I = tanh\left(\frac{\Delta t}{T}J\right) \tag{4.57}$$

Argument $\frac{\Delta t}{T} J = \frac{\delta q}{|Q|} \frac{\Delta t}{T}$ can be rewritten as $\frac{\frac{\delta q}{T}}{\frac{|Q|}{\Delta t}}$ and represents the ratio of executing order flow to market flow. Therefore, we can express execution imbalance through the ratio of the flows as:

$$I = tanh\left(\frac{\delta v}{2V}\right) \tag{4.58}$$

Execution probabilities caused by order flow imbalance can be determined as

$$p_a = \frac{1}{2} + \frac{1}{2} tanh\left(\frac{\delta v}{2V}\right) \tag{4.59}$$

$$p_b = \frac{1}{2} - \frac{1}{2} tanh\left(\frac{\delta v}{2V}\right) \tag{4.60}$$

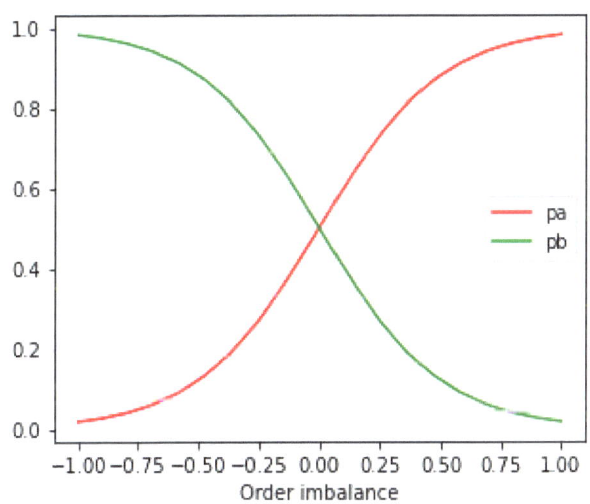

FIGURE 4.28 *Dependence of execution probabilities on order imbalance.*

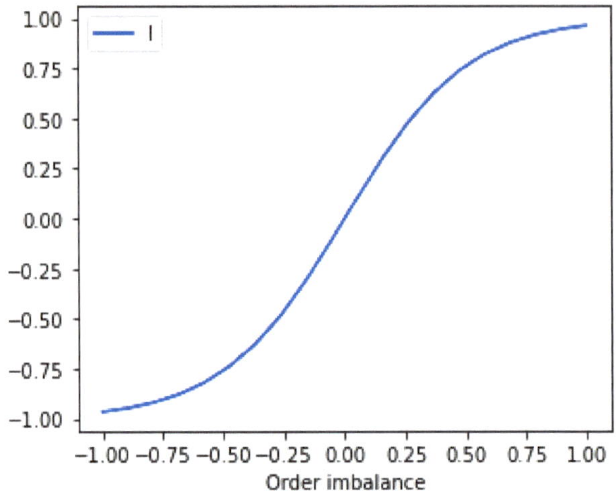

FIGURE 4.29 *Dependence of execution imbalance on order imbalance.*

4.11 Price impact in ensemble theory

Eq. (4.41) can be reversed to express impact of a newly submitted order on price. If a security is traded at fair value and has a balanced order book, then submitting an order q will create imbalance $I(q)$. This imbalance can be offset by a hypothetical order from the opposite side introducing imbalance I' such that $I + I' = 0$. Since such order is missing, the only way for the market to come back to balance is through shifting the price by amount:

$$\delta s = \Delta s \, arctanh(I) \tag{4.61}$$

Inserting Eq. (4.58) here, we get:

$$\delta s = \Delta s \frac{v}{2V} \tag{4.62}$$

Taking into account that $\frac{\Delta s}{2V} = \frac{s}{\theta}$ and $q = vT$, price impact equals

$$\delta s = s\frac{q}{\rho} \tag{4.63}$$

which means that the price shift is equal to order's penetration depth corresponding to order density in the book. If Eq. (4.45) were used, then the impact would be:

$$\delta s = \frac{s_a - s_b}{2}\frac{q}{|q|} + s\frac{q}{\rho} \tag{4.64}$$

Given the fact that order may execute anywhere within the spread, the impact is:

$$\delta s = \frac{s_a - s_b}{2} z + s\frac{q}{\rho}$$

where z is a random number.

4.12 Simplistic model of execution probabilities

In a balanced market the execution probabilities for bid and ask prices are 50/50. This split refers to all trading volume. Writing the split in terms of volume, we have

$$p_a = \frac{1}{2} \equiv \frac{V}{2V} \tag{4.65}$$

If order flow $v > 0$ is added for execution, then quantity executing at ask price will be $V + v$, and it will remain V if $v < 0$. We can write this through Heaviside step function $H(x)$ as $V + vH(v)$. Total volume is $2V + |v|$ either way. Therefore, execution probability p_a is:

$$p_a = \frac{V + vH(v)}{2V + |v|} \tag{4.66}$$

Using the identity

$$H(v) = \frac{1}{2} + \frac{1}{2}\frac{|v|}{v}$$

it can be rewritten as

$$p_a = \frac{1}{2}\left[1 + \frac{v}{2V + |v|}\right] \quad (4.67)$$

Likewise,

$$p_b = \frac{1}{2}\left[1 - \frac{v}{2V + |v|}\right] \quad (4.68)$$

These probabilities display the already familiar qualitative behavior shown in FIGURE 4.30. For $\frac{|v|}{V} \ll 1$ they agree with Eqs. (4.59) and (4.60).

Execution imbalance in this simple model equals

$$I = \frac{v}{2V + |v|} \quad (4.69)$$

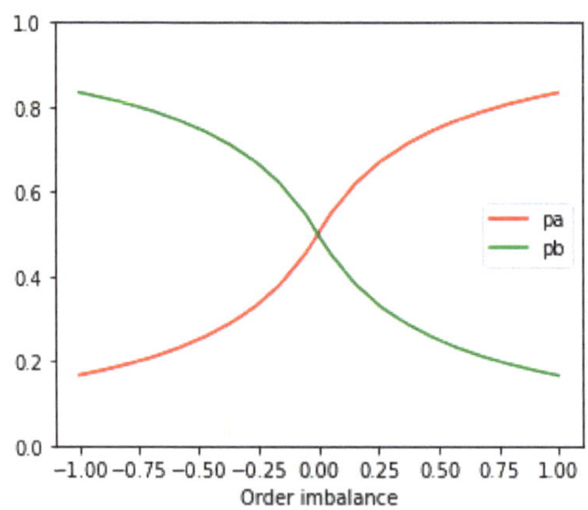

FIGURE 4.30 *Dependence of execution probabilities on order flow.*

APPENDIX A. Ensemble measures (with critical event)

1-minute ensemble volatility (e-VOL index), correlations (e-CORR index) and drawdown for S&P 500 index. Volatility is in basis points.

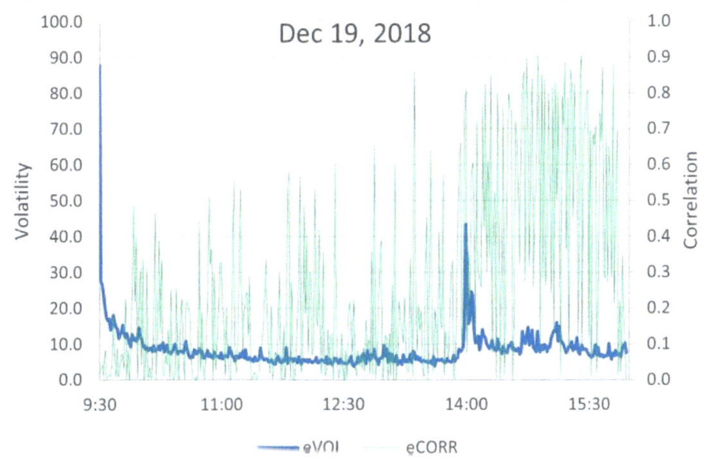

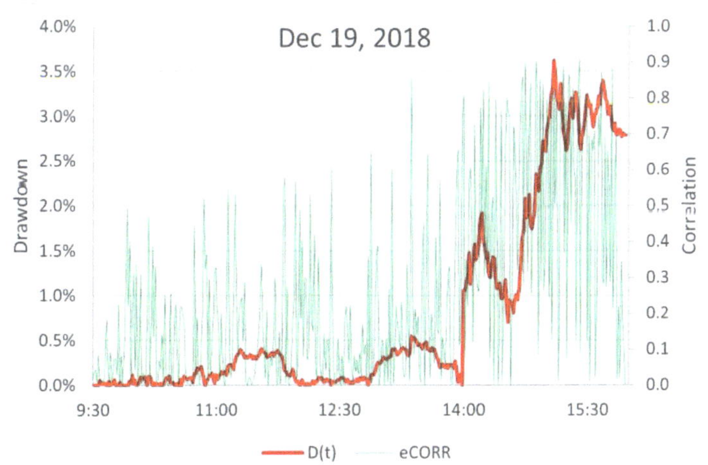

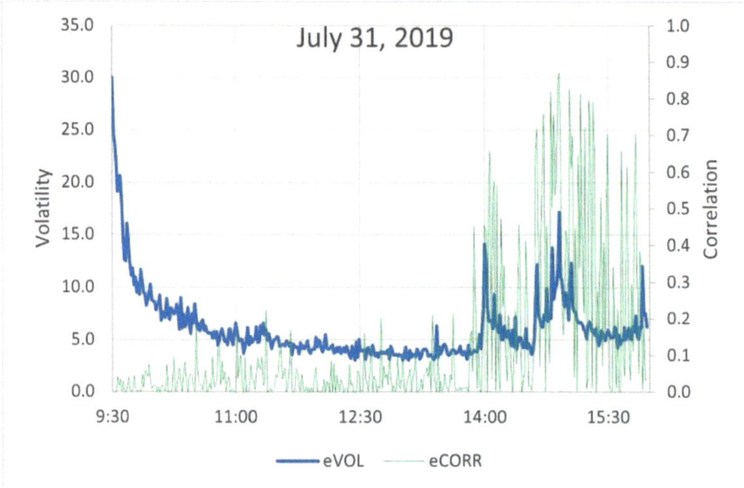

July 31, 2019
Volatility
Correlation
35.0
30.0
25.0
20.0
15.0
10.0
5.0
0.0
1.0
0.9
0.8
0.7
0.6
0.5
0.4
0.3
0.2
0.1
0.0
9:30
11:00
12:30
14:00
15:30
eVOL
eCORR

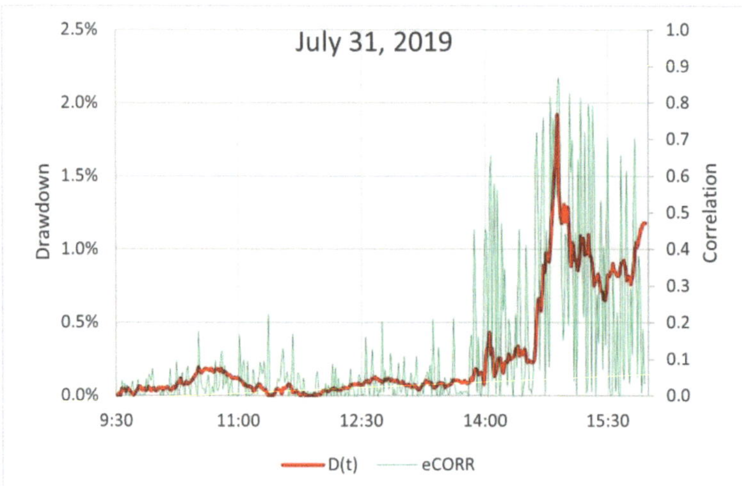

July 31, 2019
Drawdown
Correlation
2.5%
2.0%
1.5%
1.0%
0.5%
0.0%
1.0
0.9
0.8
0.7
0.6
0.5
0.4
0.3
0.2
0.1
0.0
9:30
11:00
12:30
14:00
15:30
D(t)
eCORR

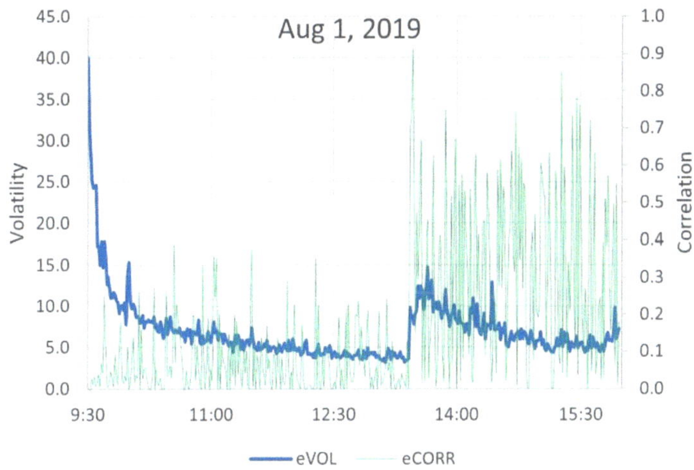
Aug 1, 2019
Volatility
Correlation
45.0
40.0
35.0
30.0
25.0
20.0
15.0
10.0
5.0
0.0
1.0
0.9
0.8
0.7
0.6
0.5
0.4
0.3
0.2
0.1
0.0
9:30
11:00
12:30
14:00
15:30
eVOL
eCORR

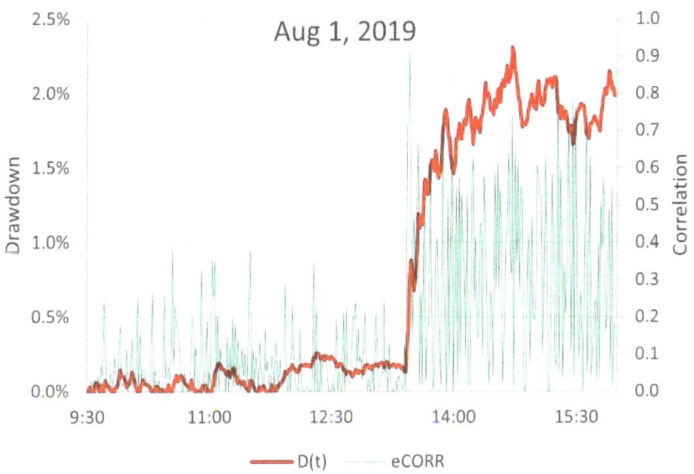
Aug 1, 2019
Drawdown
Correlation
2.5%
2.0%
1.5%
1.0%
0.5%
0.0%
1.0
0.9
0.8
0.7
0.6
0.5
0.4
0.3
0.2
0.1
0.0
9:30
11:00
12:30
14:00
15:30
D(t)
eCORR

APPENDIX B. Ensemble measures (no critical events)

1-minute ensemble volatility (e-VOL index), correlations (e-CORR index) and drawdown for S&P 500 index. Volatility is in basis points.

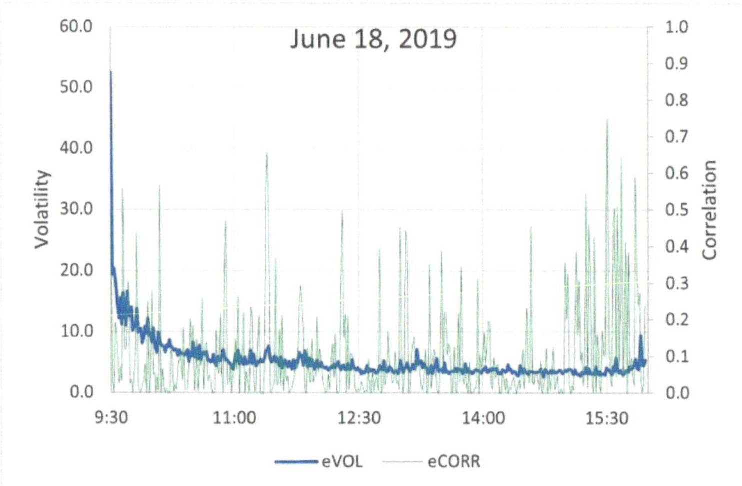

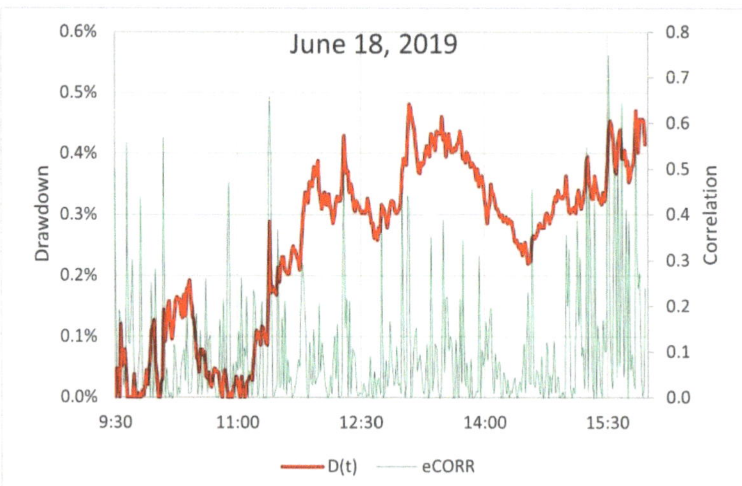

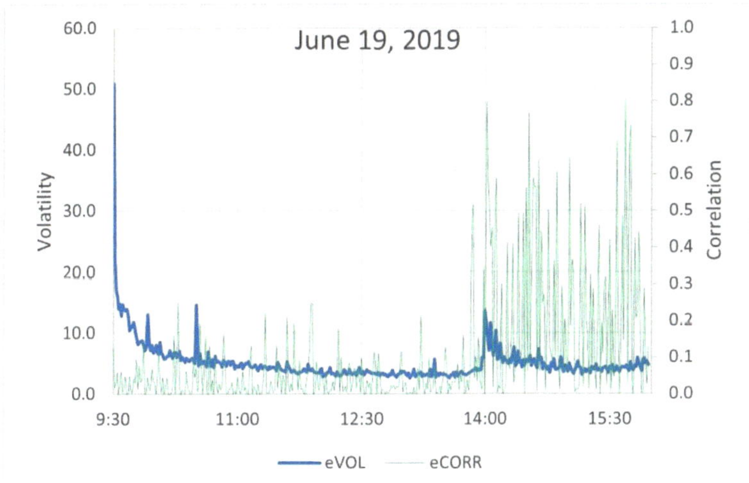
June 19, 2019
Volatility
Correlation
60.0
50.0
40.0
30.0
20.0
10.0
0.0
1.0
0.9
0.8
0.7
0.6
0.5
0.4
0.3
0.2
0.1
0.0
9:30
11:00
12:30
14:00
15:30
eVOL
eCORR

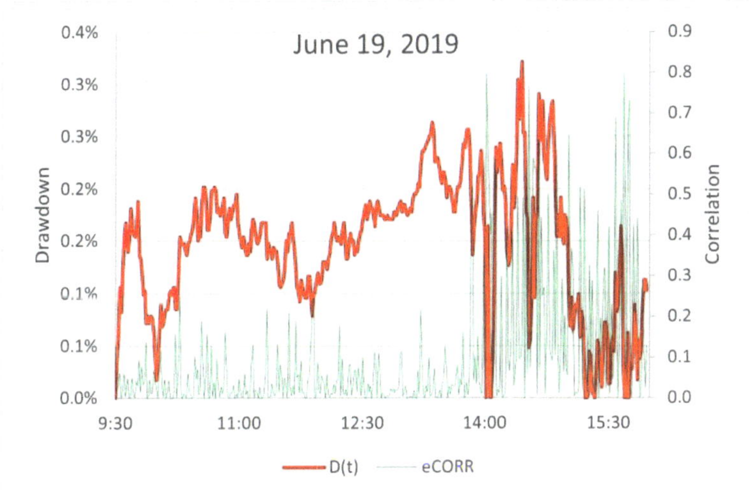
June 19, 2019
Drawdown
Correlation
0.4%
0.3%
0.3%
0.2%
0.2%
0.1%
0.1%
0.0%
0.9
0.8
0.7
0.6
0.5
0.4
0.3
0.2
0.1
0.0
9:30
11:00
12:30
14:00
15:30
D(t)
eCORR

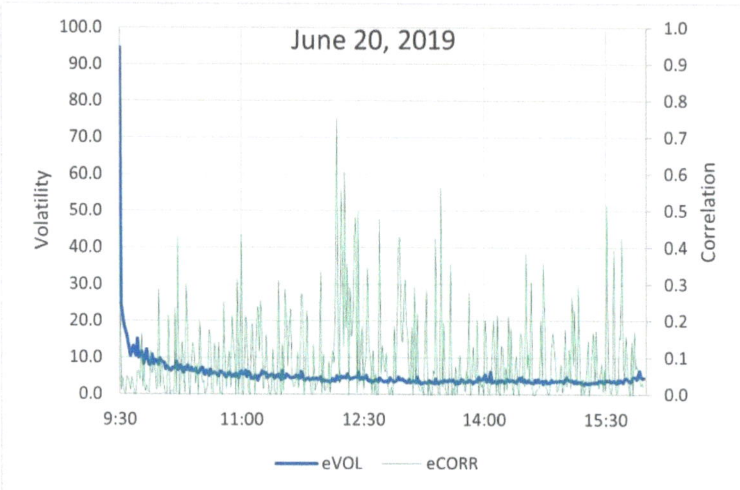
June 20, 2019
Volatility
Correlation
100.0
90.0
80.0
70.0
60.0
50.0
40.0
30.0
20.0
10.0
0.0
1.0
0.9
0.8
0.7
0.6
0.5
0.4
0.3
0.2
0.1
0.0
9:30
11:00
12:30
14:00
15:30
eVOL
eCORR

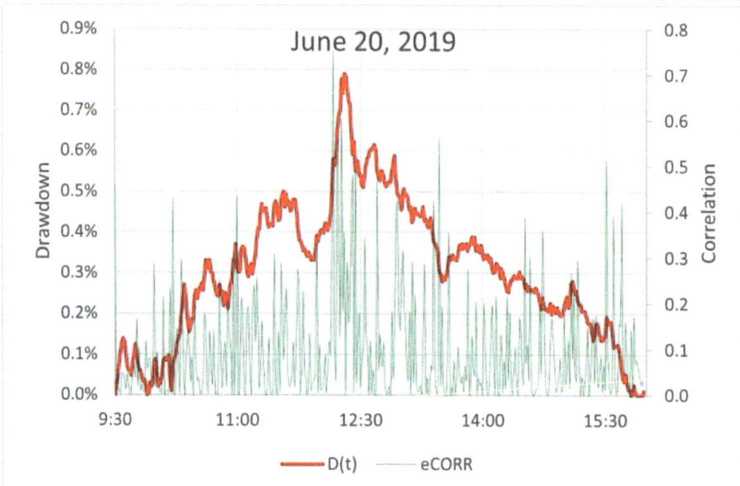
June 20, 2019
Drawdown
Correlation
0.9%
0.8%
0.7%
0.6%
0.5%
0.4%
0.3%
0.2%
0.1%
0.0%
0.8
0.7
0.6
0.5
0.4
0.3
0.2
0.1
0.0
9:30
11:00
12:30
14:00
15:30
D(t)
eCORR

APPENDIX C. Phase diagrams on days with critical events

Correlation levels vs. drawdown, 1-minute ensemble volatility (e-VOL index) and trading volume for S&P 500 index on days with substantial impact. Volatility is in basis points.

December 19, 2018

Volatility

Drawdown	0	5	10	25	50	100
0.00%	10%	17%	12%	4%	2%	
0.25%	12%	16%				
0.50%		44%				
0.75%		41%	75%			
1.00%		43%	43%	44%		
1.50%		40%	60%			
2.00%		50%	57%			
3.00%		49%	59%			
4.00%						

December 19, 2018

Volume

Drawdown	0	5000	10000	15000	20000	40000
0.00%	15%	16%				
0.25%	15%					
0.50%	36%		76%			
0.75%		44%				
1.00%	29%	50%	33%	33%	81%	
1.50%		55%	46%	86%		
2.00%		49%	48%	64%		
3.00%		44%	56%	66%	70%	
4.00%						

July 31, 2019
Volatility

Drawdown	0	5	10	25	50	100
0.00%	4%	7%	3%	0%		
0.25%	9%	25%	43%			
0.50%	14%	38%	72%			
0.75%	19%	30%	61%			
1.00%		35%	46%			
1.25%		40%	57%			
1.50%		54%	58%			
1.75%			84%			
2.00%						

July 31, 2019
Volume

Drawdown	0	5000	10000	15000	20000	40000
0.00%	5%	52%				
0.25%	15%	41%				
0.50%	31%	53%	39%			
0.75%	23%	36%	69%			
1.00%	28%	37%	44%			
1.25%	19%	33%	76%		57%	
1.50%			54%	58%		
1.75%				84%		
2.00%						

August 1, 2019
Volatility

Drawdown	0	5	10	25	50	100
0.00%	6%	8%	5%	1%		
0.25%	0%	55%				
0.50%		47%	29%			
0.75%		54%	52%			
1.00%		22%	24%			
1.50%	19%	34%	26%			
2.00%	21%	37%	9%			
3.00%						
4.00%						

August 1, 2019
Volume

Drawdown	0	5000	10000	15000	20000	40000
0.00%	7%	21%				
0.25%	11%	72%		71%		
0.50%		38%				
0.75%		54%	53%			
1.00%	0%	29%	2%			
1.50%	29%	35%	9%			
2.00%	30%	45%	3%		9%	
3.00%						
4.00%						

APPENDIX D. Phase diagrams on days without critical events

Correlation levels vs. drawdown, 1-minute ensemble volatility (e-VOL index) and trading volume for S&P 500 index on consecutive days without substantial impact. Volatility is in basis points.

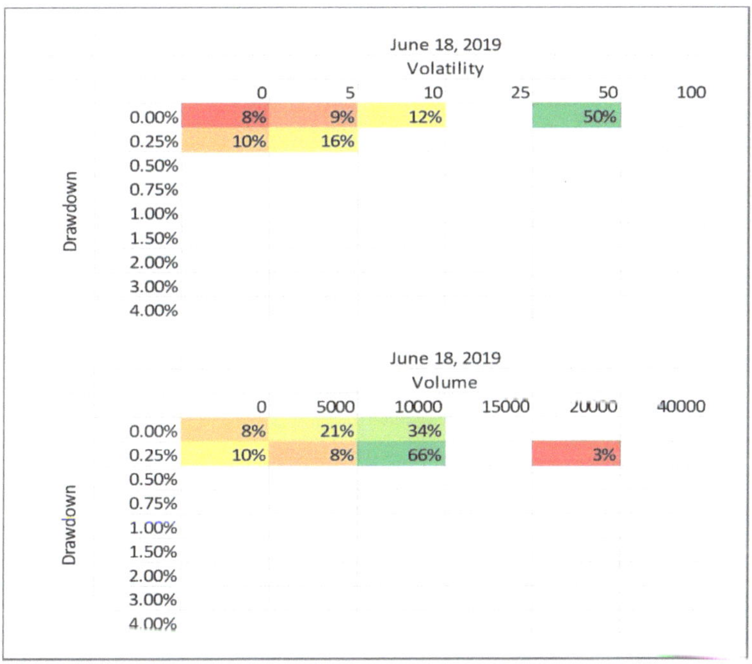

June 18, 2019

Drawdown \ Volatility	0	5	10	25	50	100
0.00%	8%	9%	12%		50%	
0.25%	10%	16%				
0.50%						
0.75%						
1.00%						
1.50%						
2.00%						
3.00%						
4.00%						

June 18, 2019

Drawdown \ Volume	0	5000	10000	15000	20000	40000
0.00%	8%	21%	34%			
0.25%	10%	8%	66%		3%	
0.50%						
0.75%						
1.00%						
1.50%						
2.00%						
3.00%						
4.00%						

June 19, 2019
Volatility

Drawdown	0	5	10	25	50	100
0.00%	9%	12%	10%		13%	
0.25%	6%	32%				
0.50%						
0.75%						
1.00%						
1.50%						
2.00%						
3.00%						
4.00%						

June 19, 2019
Volume

Drawdown	0	5000	10000	15000	20000	40000
0.00%	9%	36%	37%			
0.25%	16%					
0.50%						
0.75%						
1.00%						
1.50%						
2.00%						
3.00%						
4.00%						

June 20, 2019
Volatility

Drawdown	0	5	10	25	50	100
0.00%	9%	9%	3%		46%	
0.25%	9%	10%				
0.50%	14%	45%				
0.75%	30%	36%				
1.00%						
1.50%						
2.00%						
3.00%						
4.00%						

June 20, 2019
Volume

Drawdown	0	5000	10000	15000	20000	40000
0.00%	8%	10%	3%		51%	
0.25%	9%					
0.50%	16%					
0.75%	31%					
1.00%						
1.50%						
2.00%						
3.00%						
4.00%						

Bibliography and References

C. Caginalp, G. Caginalp. 2019. *"Price equations with symmetric supply/demand; implications for fat tails".* s.l. : Economics Letters, 176, 2019.

C. Plott, K. Pogorelsky. 2017. *"Call Market Experiments: Efficiency and Price Discovery Through Multiple Calls and Emergent Newton Adjustments".* s.l. : American Economic Journal: Microeconomics , 2017.

Derman, E. 2011. *"Models.Behaving.Badly.: Why Confusing Illusion with Reality Can Lead to Disaster, on Wall Street and in Life".* s.l. : Free Press, 2011.

Dirac, P. A. M. 1982. *"The Principles of Quantum Mechanics".* s.l. : Oxford Univ Pr., 1982.

FASB. 2010. *Statement of Financial Accounting Standards 157: Fair Value Measurements.* 2010.

G.G. Malinetsky, A.B. Potapov. 2000. *"Modern Problems of Nonlinear Dynamics".* Moscow : s.n., 2000.

H. Kumar, J. Sarkissian. 2019. *"Morning Bid-Ask Spread Relaxation: Kinetic and Quantum Frameworks".* s.l. : 8th Annual Conference on High-Frequency Finance and Analytics, Stevens Institute of Technology, 2019.

Halperin, I. and Dixon, M. . 2020. *"Quantum Equilibrium-Disequilibrium: Asset price dynamics, symmetry breaking, and defaults as dissipative instantons".* s.l. : Physica A: Statistical Mechanics and its Applications, 2020. p. 537.

Hull, J. 2015. *"Risk Management and Financial Institutions", Wiley, 4th edition.* s.l. : Wiley, 2015.

ISDA. 2016. *2016 Credit Support Annex for Variation Margin.* 2016.

J. Hirshleifer, A. Glazer, D. Hirshleifer. 2005. *"Price theory and Applications: Decisions, Markets, and Information", 7th edition,.* s.l. : Cambridge University Press, 2005.

J. Sarkissian, J. Sebold. 2018. *"Measurement of Current Market Correlations Based on Ensemble Statistics".* 2018. https://papers.ssrn.com/sol3/papers.cfm?abstract_id=3308669.

J.P. Bouchaud, M. Potters. 2004. *Theory of Financial Risk and Derivative Pricing: From Statistical Physics to Risk Management.* s.l. : Cambridge University Press, 2004.

Jorion, P. 2006. *"Value at Risk: The New Benchmark for Managing Financial Risk", 3rd edition.* s.l. : McGraw-Hill Education, 2006.

L.D. Landau, E.M. Lifshitz. 1981. *"Quantum Mechanics: Non-Relativistic Theory". Vol. 3, 3rd ed.* s.l. : Butterworth-Heinemann, 1981.

—. 1980. *"Statistical Physics". Vol. 5, 3rd ed.* s.l. : Butterworth-Heinemann, 1980.

London, J. 2004. *Modeling Derivatives in C++.* s.l. : Wiley, 2004.

Nastasiuk, V.A. 2014. *"Emergent quantum mechanics of finances".* s.l. : Physica A: Statistical Mechanics and Its Applications. Elsevier BV., 2014. pp. 148-154. Vol. 403.

—. 2015. *"Fisher information and quantum potential well model for finance".* s.l. : Physics Letters A. Elsevier BV., 2015. pp. 1998-2000. Vol. 379 (36).

Orrell, D. 2020. *"A quantum model of supply and demand".* s.l. : Physica A: statistical Mechanics and its Applications, 2020. Vol. 539.

Salov, V. 2019. *"Explosions" of Corn Futures Prices.* s.l. : Available at SSRN: https://ssrn.com/abstract=3479685, 2019.

Sarkissian, J. 2016. *"Express measurement of market volatility using ergodicity concept".* 2016. https://papers.ssrn.com/sol3/papers.cfm?abstract_id=2812353.

—. 2020. *"Statistical Mechanics of Price Stabilization and Destabilization in Financial Markets".* 2020. https://papers.ssrn.com/sol3/papers.cfm?abstract_id=3514329.

—. 2013. *"Coupled mode theory of stock price formation".* 2013. https://arxiv.org/abs/1312.4622.

—. 2017. *"Dynamic Coupling in Order Book: Quantum Framework".* 2017. https://ssrn.com/abstract=3095073.

—. 2016. *"Option pricing under quantum theory of securities price formation".* 2016. http://papers.ssrn.com/sol3/papers.cfm?abstract_id=2848014.

—. 2019. *"Price Measurement in Financial Markets and Quantum Coupled-Wave Model of Price Dynamics".* s.l. : American Physical Society Annual Meeting, 2019.

—. 2019. *"Price Measurement in Financial Markets and Quantum Coupled-Wave Theory of Price Formation".* s.l. : 8th Annual Conference on High-Frequency Finance and Analytics at Stevens Institute of Technology, 2019.

—. 2016. *"Quantum theory of securities price formation in financial markets".* 2016. http://ssrn.com/abstract=2765298.

—. 2016. *"Risk Valuation for Securities with Limited Liquidity".* 2016. https://papers.ssrn.com/sol3/papers2.cfm?abstract_id=2891669.

—. **2019.** *"Self-Controlled Phase Transitions During Market Crashes and Price Corrections".* 2019. https://ssrn.com/abstract=3512362.

—. **2016.** *"Spread, volatility, and volume relationship in financial markets and market maker's profit optimization".* 2016. http://ssrn.com/abstract=2799798.

—. **2020.** *Quantum coupled-wave theory of price formation in financial markets: price measurement, dynamics and ergodicity.* s.l. : Physica A, 2020. Vol. 554.

Sinai, I.G. 1977. *"Introduction to Ergodic Theory", Princeton University Press; First Edition.* 1977.

Stöckmann, H-J. 2007. *"Quantum Chaos: An Introduction", First Edition.* s.l. : Cambridge University Press, 2007.

Stoikov, S. 2018. *"The micro-price: a high-frequency estimator of future prices".* s.l. : Quantitative Finance, 2018.

Taleb, N.N. 2010. *"The Black Swan: Second Edition".* s.l. : Random House, 2010.

—. **2005.** *Fooled by Randomness: The Hidden Role of Chance in Life and in the Markets.* s.l. : Random House, 2005.

Voit, J. 2001. *The Statistical Mechanics of Financial Markets (Theoretical and Mathematical Physics).* s.l. : Springer, 2001.

Zel'dovich, B. Ya., Pilipetsky, N.F. and Sukhov, V.V. 2013. *"Principles of Phase Conjugation".* s.l. : Springer-Verlag Berlin and Heidelberg GmbH & Co. KG, 2013.

Zel'dovich, B.Ya. 2008. *"Impedance and parametric excitation of oscillators".* s.l. : Phys. Usp., 2008. pp. 465–484. Vol. 51.

Zel'dovich, B.Ya. and Soileau, M.J. . 2004. *"Bi-frequency pendulum on a rotary platform: modeling various optical phenomena"*. s.l. : Phys. Usp., 2004. pp. 1239–1255. Vol. 47.

Index

About the author

Jack Sarkissian (PhD) is an asset manager and physicist, former banker, currently overseeing automated trading as the Managing Partner of Algostox Trading. He was the Chief Investment Officer for EG Capital Partners managing $3 billion in assets, and prior to that ran a risk management department and held senior quantitative positions in investment banks.

Jack applies methods of theoretical physics to reveal fundamental properties of financial markets. His research addresses problems he had/has to deal with himself as an institutional asset manager and trader, including trade execution, order book dynamics, block trade pricing, risk assessment of illiquid securities, OTC derivatives pricing, market microstructure, and ensemble properties of financial markets.

Jack's research led him to the formulation of Quantum Theory of Price Formation, which not only takes our understanding of financial processes to a new level, but also opens new paths for trade execution modeling, securities pricing, and risk management.

www.ingramcontent.com/pod-product-compliance
Lightning Source LLC
LaVergne TN
LVHW052148210926
848482LV00018B/12
9780578796376